Proceedings

Workshop on Human Motion

7-8 December 2000

Austin, Texas

Sponsored by

IEEE Computer Society

IEEE Computer Society Technical Committee on
Pattern Analysis and Machine Intelligence

Los Alamitos, California

Washington   •   Brussels   •   Tokyo

IEEE Computer Society Order Number PR00939
ISBN 0-7695-0939-8
ISBN 0-7695-0940-1 (case)
ISBN 0-7695-0941-X (microfiche)
Library of Congress Number 00-110409

Additional copies may be ordered from:

IEEE Computer Society	IEEE Service Center	IEEE Computer Society
Customer Service Center	445 Hoes Lane	Asia/Pacific Office
10662 Los Vaqueros Circle	P.O. Box 1331	Watanabe Bldg., 1-4-2
P.O. Box 3014	Piscataway, NJ 08855-1331	Minami-Aoyama
Los Alamitos, CA 90720-1314	Tel: + 1 732 981 0060	Minato-ku, Tokyo 107-0062
Tel: + 1 714 821 8380	Fax: + 1 732 981 9667	JAPAN
Fax: + 1 714 821 4641	http://shop.ieee.org/store/	Tel: + 81 3 3408 3118
http://computer.org/	customer-service@ieee.org	Fax: + 81 3 3408 3553
csbooks@computer.org		tokyo.ofc@computer.org

Editorial production by Danielle C. Young

Cover art production by Joe Daigle/Studio Productions

Printed in the United States of America by Applied Digital Imaging

IEEE
COMPUTER
SOCIETY

Table of Contents 2000

Workshop on Human Motion (HUMO 2000)

Session 4: Motion Estimation & Recognition II

Session 5: Multiple Persons or Multiple Cameras I

Session 6: Multiple Persons or Multiple Cameras II

Session 7: Modeling/Counting

Session 8: Interaction and Shape Estimation

Foreword

It is our pleasure to present these Proceedings of the IEEE Computer Society Workshop on Human Motion (HUMO 2000). The proceedings consists of 23 submitted papers. Profs. Aaron Bobick and Larry Davis exercised considerable care in the review of submitted manuscripts with the help of a cadre of reviewers. The program will also include three invited papers.

This workshop follows the earlier workshops in Austin (Motion of Non-rigid and Articulated Objects, 1994) and Puerto Rico (Nonrigid and Articulated Motion, 1997). There have been several other workshops on human motion, including those on surveillance (in Bombay, India (1998), Fort Collins, Colorado (1999) and Dublin, Ireland (2000)), the HuMAnS 2000 Workshop in Hilton Head, NC (June 2000), and the First International Workshop on Articulated Motion and Deformable Objects, Palma de Mallorca, Spain (September 2000). These workshops, as well as earlier ones, have followed the evolution of motion research, beginning with the study of the motion of rigid objects in the 1970's through the present interest in the motion of non-rigid objects. The study of human motion is a fascinating subject with the attendant complexities of both non-rigidity and occlusion.

We hope that the workshop papers presented here capture an important slice of today's research and give an indication of the future directions in human motion understanding. The final versions of some of the papers were received late for the review process, indicating the results are hot off the press.

We wish to express our appreciation to all of the authors who submitted papers to the workshop, to the program co-chairman, Profs. Aaron Bobick and Larry Davis, and to our program committee members who reviewed the papers in a short time. Local arrangements and administrative assistance was provided by Mrs. D. Prather. Finally, we wish to thank the workshop participants for their interest and support.

J. K. Aggarwal

Alex Pentland

Workshop General Chairmen

Program Committee

Program Co-Chairs

Aaron Bobick, Georgia Institute of Technology
Larry S. Davis, University of Maryland

Committee Members

Yiannis Aloimonos
Department of Computer Science
University of Maryland

Nicholas Ayache
INRIA Sophia-Antipolis

Michael Black
Department of Computer Science
Brown University

Andrew Blake
Microsoft Research
Cambridge

Christoph Bregler
Computer Science Department
Stanford University

Trevor Darrell
Media Laboratory
Massachusetts Institute of Technology

James W. Davis
Computer and Information Science Department
Ohio State University

Irfan Essa
College of Computing
Georgia Institute of Technology

Dmitry B Goldgof
Department of Computer Science & Electrical Engineering
University of South Florida

Thomas S. Huang
Beckman Institute for Advanced Science and Technology
University of Illinois, Urbana-Champaign

Ionnis A. Kakadiaris
Department of Computer Science
University of Houston

Takeo Kanade
Robotics Institute
Carnegie Mellon University

Steve Maybank
Department of Computer Science
University of Reading

Dmitris Metaxas
Department of Computer & Information Science
University of Pennsylvania

Pietro Perona
Department of Electrical Engineering
California Institute of Technology

Mubarak Shah
School of Elec. Engineering & Computer Science
University of Central Florida

Yoshiaki Shirai
Department of Computer-Controlled Mechanical Systems
Osaka University

Mohan Trivedi
Department of Electrical and Computer Engineering
University of California San Diego

Christopher R. Wren
Perceptive Network Technologies, Inc.
Waltham MA

Session 1

Direction & Tracking I

Phase in Model-Free Perception of Gait

Jeffrey E. Boyd
Dept. of Computer Science
University of Calgary
Calgary AB T2N 1N4
boyd@cpsc.ucalgary.ca

James J. Little
Department of Computer Science
University of British Columbia
Vancouver, B.C., Canada V6T 1Z4
little@cs.ubc.ca

Abstract

Variations in human gaits are manifest in the timing of the many combined motions in the gait. In periodic systems, such as a gait, timing reduces to phase. Therefore, in order to capture the important information in the timing patterns in a gait, one must consider phase. Gaits vary for several reasons, including different builds, moods of individuals, fatigue, and injury. We investigate the relationship between the model-free shape-of-motion phase analysis and a subjective description of gait, such as a normal gait versus a tired gait or a shuffle, by analyzing several gait image sequences that differ subjectively. A simple model based on a phasor representation of gait motion relates the pendulum-like motion of limbs to shape-of-motion features. Our ultimate goal is to develop a gait feature space that can be partitioned according to subjective perception of gait. Gait features that vary with subjective changes in gait lead in this direction.

1. Introduction

The perception of gait has long been of interest to researchers in many disciplines. A recurring theme in this area is the importance of timing of the components of a gait. For example, Murray et al. [12] show a detailed analysis of the gaits for several *normal* humans. The analysis focuses on the timing of joint and limb movements throughout the cycle of the gait. Of particular interest are the measurements of *forward displacement* which show the sinusoidal variations in a subjects forward progress as a function of position in the gait cycle. In subsequent work, Murray [11] shows that the these sinusoidal variations change in phase for subjects that have abnormal gaits due to some physical affliction. The changes in the gaits are manifest in the timing patterns.

In periodic systems, such as a human gait, timing reduces to phase. That is, temporal relationships reduce to locating the points in the period where events occur. Once a periodic reference event is chosen, e.g., a left foot fall, then the timing of all components of the gait can be described as either leading or following the reference by a specific fraction of the period. Therefore, in order to capture the important information in the timing patterns in a gait, one must consider phase.

Bertenthal and Pinto [3] identify the following three important properties in the perception of human gaits.

Frequency entrainment. The requirement that the various components of the gait must share a common frequency.

Phase locking. The requirement that the phase relationships among the components of the gait remain approximately constant. The lock varies for different types of locomotion such as walking versus running.

Periodic attractors. The gait must be a stable solution to an equation of motion. Each joint in a gait forms a stable phase plane trajectory, or limit cycle.

They state that "*...the clear conclusion that emerges is that discrete forms of relative phase ... are necessary for the perception of biomechanical motions.*" Here *discrete* refers to phase-locked modes. Furthermore, Bertenthal and Pinto suggest perception and synthesis of gait are related, "*that the perception of human movements is similar to the production of these movements in that the same order parameters or collective variables used to describe the control and coordination of human movements can be used to recognize and discriminate different movement patterns.*"

If this is true, then systems that synthesize gaits can provide clues to gait perception. Laszlo et al. [8] describe a limit-cycle control system for animation of walking figures. Varying phases in the limit cycles alters the type of gait. It also alters the perceived gait. Unuma et al. [14] describe a Fourier-based system for gait synthesis. Their system derives Fourier coefficients, consisting of magnitude and phase, for the periodic joint angles in a gait. One set of coefficients represents a *normal* gait while another set represents a variation such as a *tired* gait. By interpolating or extrapolating both the magnitude and phase coefficients they are able to blend subjectively different gaits and realize a continuum of gaits. The phase component of the Fourier series plays a significant role in the synthesis of the varying gaits. In related work, Amaya et al. [1] synthesize motions that convey emotion by rescaling the timing and amplitude of a set of non-periodic motions based on measurements of normal and expressive examples.

Other than the *forward displacement* measured by Murray [11], these analyses of gait are based on a kinematic model. They refer to the motion of individual joints or limbs. Any perception of the gait must then reconcile the many measurements for the body parts. In contrast, Cutler and Davis [6] describe a method to detect periodic motions without a kinematic model. Little and Boyd [9] also use a *shape-of-motion* model-free approach that combines motion over an entire moving figure into a set of scalars that describes the shape of the region moving in an image. Phase differences in the variation of the scalars are able to distinguish among gaits of different human subjects. They suggest that the scalars are able to recognize gaits because they represent a spatial average of the phase information at individual points in the moving figure [4].

Gaits vary for several reasons, including different builds, moods of individuals, fatigue, and injury. In this paper we investigate the relationship between the model-free shape-of-motion phase analysis and a subjective description of gait. Subjective descriptions refer to general descriptions of gait related to perception rather than to some quantitative evaluation. Examples include such descriptors as normal, shuffle, bouncy, tired, and sneaky. Following well established techniques of using synthetic gaits to provide control in perceptual experiments [7, 3], we use computer animation techniques to generate gaits that vary only by some parameter that subjectively describes the gait. We then examine the variations in the phase features that result. Our ultimate goal is to develop a gait feature space that can be partitioned according to subjective perception of gait. Gait features that vary with subjective changes in gait lead in this direction.

2. Frequency Content of Gaits

As Bertenthal and Pinto point out [3], *frequency entrainment* is an important part of gait perception. It is necessary to lock-on to the frequency of a gait to analyze it. A difficulty immediately arises because there are many different frequencies of motion present in a gait. Which frequency do we then lock onto?

Figure 1 describes some of the sources of different frequencies in a gait. The fundamental frequency, f_0, is found in the pendulum-like oscillations of a limb, Figure 1(a). Here period of the motion goes from foot forward to back to forward again. If we consider a silhouette of the body, left and right are not distinguishable and the frequency is $2f_0$, the step rate, Figure 1(b). As a limb swings, it goes up and down twice during the period of the pendulum, thus giving

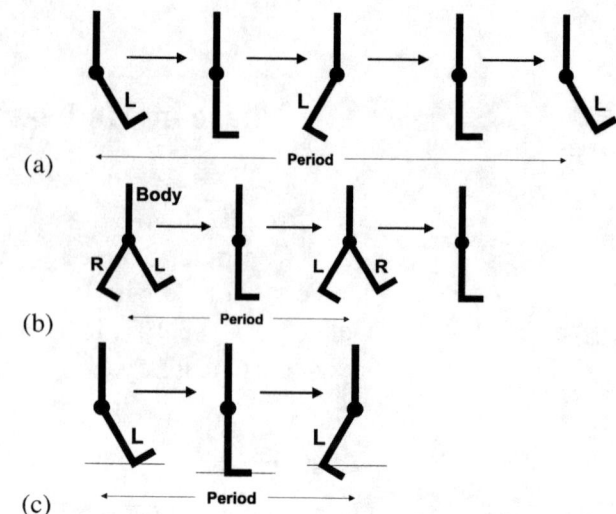

Figure 1. Stylized body and legs showing sources of different frequencies in a synthesized gait: (a) the oscillation of a swinging limb repeats periodically, e.g., left foot fall to left foot fall, (b) the silhouette of a body repeats at twice that frequency, i.e., step to step, and (c) the pendulum motion of limbs has vertical motion at twice the frequency of the limbs horizontal motion.

a vertical motion at twice the fundamental frequency, again $2f_0$. Figure 2 shows three spectra derived from the motion of selected body parts in a synthetic gait produced by Poser (see Section 4). The spectra are produced using the least-squares linear prediction method [2]. Frequencies in the spectra are normalized to the sampling frequency ($30fps$). The spectra show both f_0 and $2f_0$ frequency components. Note that an additional component at $3f_0$ is present in the chest motion, Figure 2(c). We do not know how this frequency arises.

3. Shape of Motion and Phase Features

Figure 3 illustrates the data flow through the *shape-of-motion* gait recognition system that creates the features that are used for recognition [9]. The system begins with an image sequence of $n + 1$ images featuring the frontoparallel motion of a single pedestrian walking in front of a static background, and then derives n dense optical flow images [5]. For each of these optical flow images, the system computes m characteristics that describe the shape of the motion (i.e., the spatial distribution of the flow), for example, the centroid of the moving points, and various moments of the flow distribution. Some of these are locations in the image, but all are treated as time-varying scalar values. Table 1 summarizes the scalar values used. Rearranging the scalar values forms a time series for each scalar. The system analyzes the periodic structure of these time series and

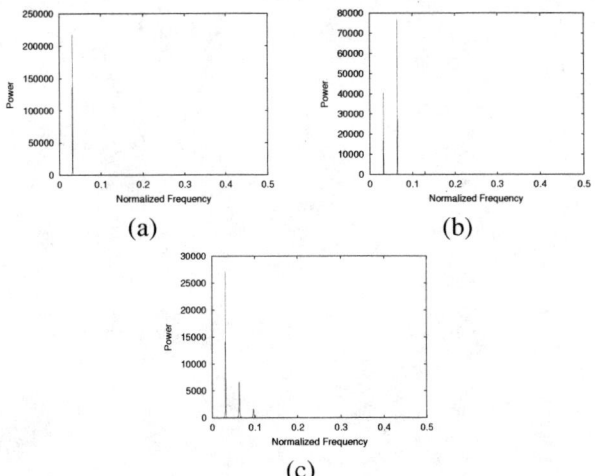

(a)   (b)

(c)

Scalar	Formula
x_c	$\sum xT / \sum T$
y_c	$\sum yT / \sum T$
x_{wc}	$\sum x\lvert(u,v)\rvert T / \sum \lvert(u,v)\rvert T$
y_{wc}	$\sum y\lvert(u,v)\rvert T / \sum \lvert(u,v)\rvert T$
x_d	$x_{wc} - x_c$
y_d	$y_{wc} - y_c$
a_c	$\lambda_{max}/\lambda_{min}$
a_{wc}	$\lambda_{max}/\lambda_{min}$
a_d	$a_{wc} - a_c$
x_{uwc}	$\sum x\lvert u\rvert T / \sum \lvert u\rvert T$
y_{uwc}	$\sum y\lvert u\rvert T / \sum \lvert u\rvert T$
x_{vwc}	$\sum x\lvert v\rvert T / \sum \lvert v\rvert T$
y_{vwc}	$\sum y\lvert v\rvert T / \sum \lvert v\rvert T$

Table 1. Summary of scalar shape-of-motion descriptors. Summations are over the entire image. u and v are the x- and y-direction optical flow values respectively. The function T segments the image. $T = 1$ for pixels that are moving and $T = 0$ for stationary pixels. λ s for a_c and a_{wc} are eigenvalues of the covariance matrix for the corresponding motion distributions.

Figure 2. Frequency spectra derived from the motion of various body parts in a gait. The horizontal motion of the right hand (a) shows the fundamental frequency of the gait, f_0. The vertical motion of the right hand (b) shows both a f_0 and a $2f_0$ component. The horizontal motion of the chest shows three components at f_0, $2f_0$ and $3f_0$.

Little and Boyd [9] demonstrated recognition rates of individual gaits among a set of gaits for six people in the 90% range. The value of the method was further supported by statistical analysis confirming significant variation of phase features with individual gaits.

Shape-of-motion recognition exploits the requirement for phase locking in gait perception. To understand the relationship between phase locking and the shape of motion we turn to *phasors*, a mathematical tool that is useful for describing sinusoids of a common frequency.

The following summary of *phasors* is adapted from Smith [13]. Consider the function $\mathbf{W}(t)$, a complex function of time, where

$$\mathbf{W}(t) = Ae^{j(\omega t + \phi)}$$

The real and imaginary parts of $\mathbf{W}(t)$ are

$$\begin{aligned} W_{real} &= A\cos(\omega t + \phi) \\ W_{imag} &= A\sin(\omega t + \phi) \end{aligned}$$

Therefore, for a sinusoidal function with amplitude A and frequency ω such as

$$v(t) = A\cos(\omega t + \phi)$$

we can say that

$$v(t) = \mathrm{Re}\{Ae^{j(\omega t + \phi)}\} = \mathrm{Re}\{Ae^{j\omega t}e^{j\phi}\} \qquad (1)$$

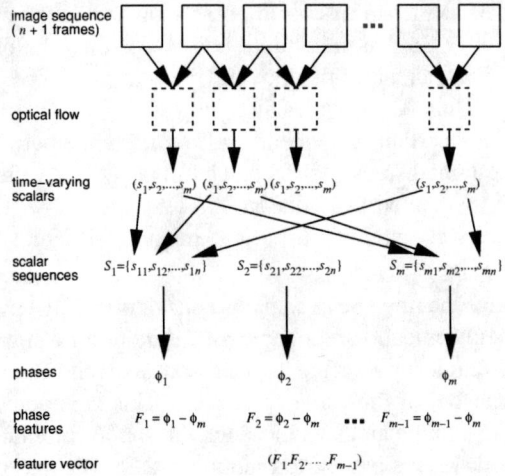

Figure 3. The structure of shape-of-motion image analysis. Each image sequence produces a vector of $m-1$ phase values used as features for recognition.

determines the frequency of the variation of each scalar. Because the system is looking at the silhouette of the moving region, this frequency is the step rate, $2f_0$. The set of time series for a view shares the same frequency but their phases vary. To make different sequences comparable, the system subtracts a reference phase, ϕ_m, derived from one of the scalars. Each image sequence is characterized by a vector, $F = (F_1, \ldots, F_{m-1})$, of $m-1$ relative phase features. The phase feature vectors are then used to recognize individuals.

5

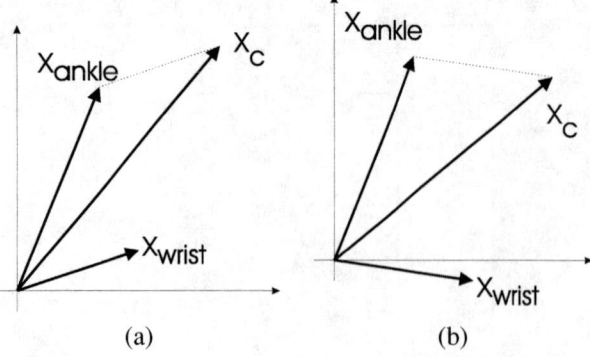

(a)   (b)

Figure 4. Phasor addition related to shape-of-motion.
The diagram on the left, (a), shows the addition of a two
phasors, one for a wrist and one for an ankle, $\mathbf{X}_{wrist}$
and $\mathbf{X}_{ankle}$. If the body parts are *phase-locked* their
sum, $\mathbf{X}_c$ will have a fixed phase. If the relative phase
changes, as in (b), then the phase of $\mathbf{X}_c$ changes too.

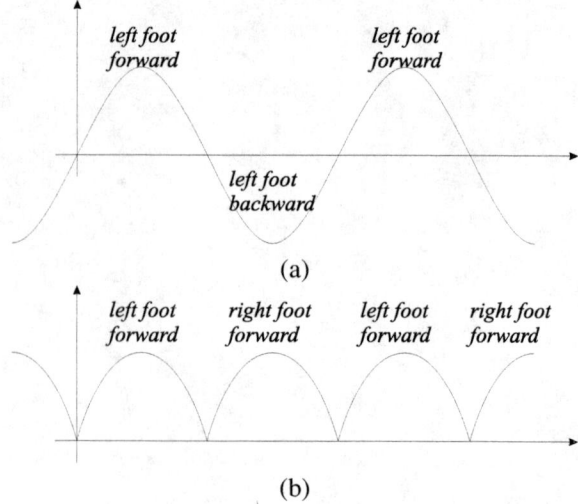

(a)

(b)

Figure 5. By looking at only moving pixels, the shape-
of-motion system effectively rectifies the motion due to
the swinging of limbs (a), and doubles the frequency of
the perceived motion when combined with the opposite
limb (b).

If we assume that we are dealing with a set of signals with
the same frequency (which is reasonable when analyzing
gaits), we can drop the $\exp(j\omega t)$ term (it acts as a rotating
reference for the signals) and deal only with the complex
number $Ae^{j\phi}$. We then represent the sinusoid $v(t)$ with the
complex *phasor* (for *phase vector*)

$$\mathbf{V} = Ae^{j\phi} = A\angle\phi.$$

A direct consequence of the phasor representation is that
we can add two sinusoidal functions simply by adding their
phasors in the complex plane, i.e., manipulations of sinusoidal functions are reduced to operations on vectors.

Now consider two body parts moving with their phases
locked in a gait. Figure 4(a) describes the situation. Two
phasors, $\mathbf{X}_{ankle}$ and $\mathbf{X}_{wrist}$ represent the sinusoidal motion
of the ankle and wrist, locked in phase at the fundamental
frequency of the gait. If one were to keep the $\exp(j\omega t)$ term
from Equation 1 then the phasors would be rotating about
the origin with angular frequency ω. The sum of the two
phasors, $\mathbf{X}_c$ is obtained by the vector addition of $\mathbf{X}_{ankle}$
and $\mathbf{X}_{wrist}$. As long as the ankle and wrist remain phase-
locked, the sum, $\mathbf{X}_c$ does not change. In Figure 4(b) the
phase-lock is altered by moving by changing the phase of
$\mathbf{X}_{wrist}$. This changes both the magnitude and phase of $\mathbf{X}_c$.

This phasor addition relates to the shape of motion as
follows. Take the example of the scalar descriptor x_c. Table
1 indicates that it is an average of x-coordinates of pixels
taken over the moving region. If we assume that each pixel
moves sinusoidally, then we can derive the sinusoidal variation of x_c by adding the phasors for all of the moving pixels,
i.e.,

$$\mathbf{X}_c = \frac{1}{|M|}\sum_{i\in M}\mathbf{X}_i \qquad (2)$$

where M is the set of moving pixels and $\mathbf{X}_i$ is the phasor
for pixel i. Just as individual body parts must be phase-
locked in a gait, all of the moving pixels must also be phase
locked. Thus $\mathbf{X}_c$ provides an aggregate view of the phase
lock for a particular sequence. The difference between a
moving-light display (MLD) and a full image is the number
of points summed in Equation 2. Therefore, an MLD can
be interpreted without attributing kinematic structure to the
points of light.

Notice that the above analysis considers only the motion
at the fundamental frequency, f_0. Addition of sinusoids
at this frequency is a linear operation and can only result
in a sinusoid of the same frequency. This contradicts the
shape-of-motion analysis in which the sum of motion over
the pixels results in a period motion at $2f_0$. The reason lies
in the non-linear thresholding operation where the shape-
of-motion systems considers only points that are moving.
At any point in a gait, one leg is nearly stationary while the
other swings forward. Thus the shape-of-motion does not
look at the full period of motion for a limb, only the forward
portion. The cycle of motion analyzed is left foot forward,
right foot forward, left foot forward and so on. This results in a doubling of frequency akin to a full-wave rectifier,
illustrated in Figure 5, in which the rectifier selects only forward motion. Equation 2 is still valid, but a single phasor,
$\mathbf{X}_i$, may correspond to two symmetrical parts on the body,
e.g., left foot and right foot, that together form a motion at
$2f_0$. If the phase angle for a single limb is ϕ, then the phase
angle for the rectified motion of it and the opposite limb is
$2\phi \pmod{360°}$. Therefore, there exists a relationship, al-

6

beit complex, between the f_0 motion of limbs and the $2f_0$ shape-of-motion features.

4. Controlled Trials

Variations in phase lock can discriminate among individual gaits as demonstrated by Little and Boyd [9]. We speculate that the variations arise for several factors, including: mood, fatigue, injury or pain, style of gait, and a person's build. While a person's build will affect their gait as a necessary consequence of kinematic structure and can be determined by recovering a kinematic model, the other factors should lead to subjectively different gaits. That is, two people with nearly identical builds that differ in their moods, fatigue, health, or style of walk, we expect to have different gaits.

This is difficult to show for gait data derived from real people. We cannot, in practice, fix one factor that affects gait and vary the others. For example, we cannot simply set a dial to adjust a persons mood. We may be able to vary fatigue, but the scale would be difficult to establish and it would definitely correlate with mood. It is unlikely that we could ever find a sample of real humans whose builds vary in some systematic manner but have identical gaits in every other respect.

Therefore, to get control over factors that affect gait, we must give up using real subjects and use synthetic gaits. The technique is well established as shown by the wide use of Cutting's canonical walker [7]. We capitalize on developments in computer animation and use the commercial product Poser 4 [10] that allows a user to produce animated poses of articulated objects, particularly humans. An important feature of Poser 4 is the ability to animate gait. A user can create a gait for a figure and adjust the gait from the *normal* gait to vary degrees of one or many subjectively different gaits. This allows us to isolate and control subjective factors of the gait. We perform three separate trials, each varying one of the following parameters: *shuffle*, *sexy-walk*, and *leg length*. Shuffle and *sexy-walk* are subjective styles of gait. The software allows the user to adjust the *amount* of the style to apply to the walk. The amount ranges from -200% to 200% with zero indicating no effect, i.e., a normal gait. For the trials reported here the amounts vary from zero to 180%. Although we do not know about the internal workings of Poser 4, we infer from using it that gaits are blended by interpolating between real, motion capture gait data, akin to the work of Unuma et al. [14], and the *amount* of a gait style is an interpolation parameter. *Leg length* is a structural parameter that is not subjective. A trial that varies leg length from normal for the Poser model to 109% of normal allows us to compare the effects of structural and subjective variations. For each trial we produce 10 gait motion sequences of about 90 frames each. We apply the shape-of-motion algorithm to the sequence to derive the

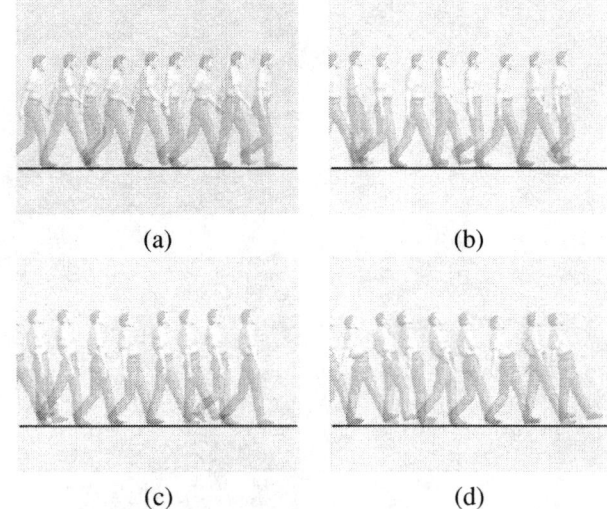

Figure 6. Examples of gait with varying degrees of *shuffle*: (a) 0%, (b) 60%, (c) 120%, and (d) 180%.

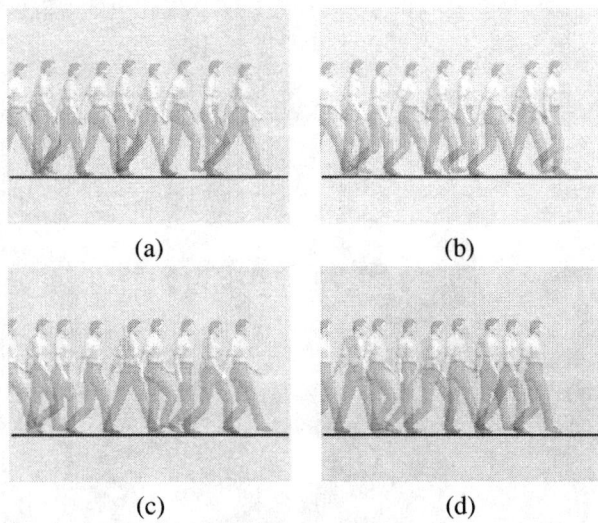

Figure 7. Examples of gait with varying degrees of *sexy-walk*: (a) 0%, (b) 60%, (c) 120%, and (d) 180%.

phase features.

Figure 6 shows four sample frames from sequences with varying *shuffle*. Figure 6(d) shows the predominant characteristic of the *shuffle* which is the legs dragging behind the torso. The images in Figure 7 show frames from sequences with varying *sexy-walk*. *Sexy-walk* is particularly hard to see in static frames. Its chief characteristic appears to be an exaggerated swinging of the hip and an unusual arm movement. Figure 8 shows frames from sequences for varying *leg length*. For each of the three factors we acquire rendered images of the gait and save Cartesian coordinates of the joint positions.

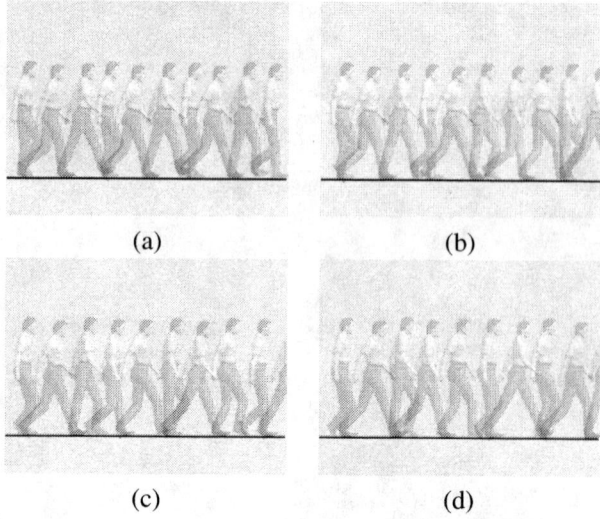

(a)   (b)

(c)   (d)

Figure 8. Examples of gait with varying leg length: (a) normal length, (b) 103%, (c) 106%, and (d) 109%.

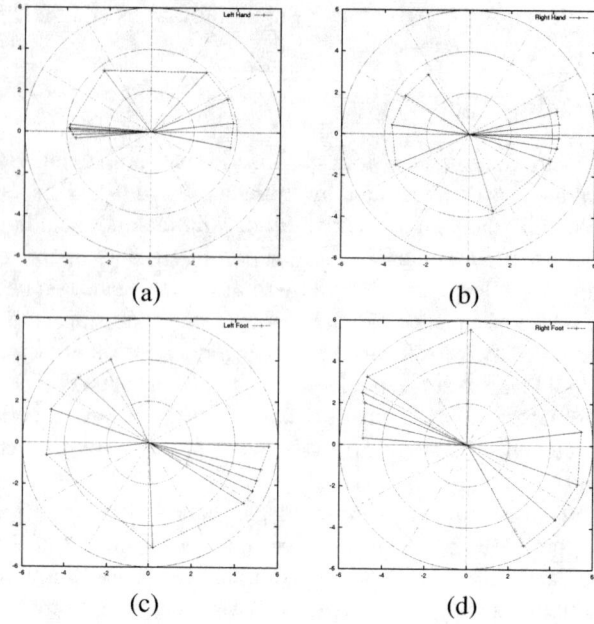

(a)   (b)

(c)   (d)

Figure 10. Phasor plots of joint movements for varying _sexy walk_ (0 to 180%): (a) left hand, (b) right hand, (c) left foot, and (d) right foot. Radial lines are the phasors. The line connecting the ends of the phasors follows the change in _sexy-walk_.

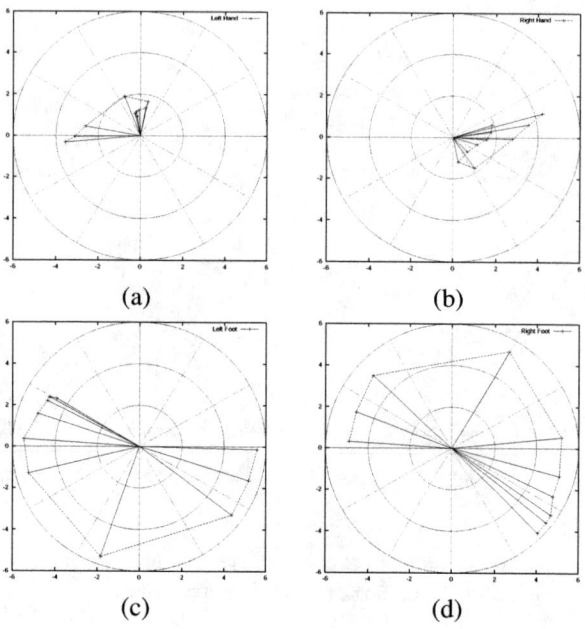

(a)   (b)

(c)   (d)

Figure 9. Phasor plots of joint movements for varying _shuffle_ (0 to 180%): (a) left hand, (b) right hand, (c) left foot, and (d) right foot. Radial lines are the phasors. The line connecting the ends of the phasors follows the change in _shuffle_.

5. Results

Joint Data: Figures 9 through 11 show polar plots of the phasors for selected body parts. The phasors are based on the fundamental frequency of the gait, and are referenced to the motion of the chest. All three plots show, as expected, that opposite limbs are nearly opposite in phase. We see that for _shuffle_ and _sexy-walk_, Figures 9 and 10, the phase variation is large. For example, the phase change in leg movement is over $180°$. The phase change for the hands is largest for _sexy-walk_ but is significant for _shuffle_ too. This demonstrates the changes in phase lock that occur with a change in style of gait. In contrast, the phase variation for changes in leg length is negligible, and although the build of the person changes significantly, the phase lock does not.

Shape of Motion: Figure 12 shows the results of shape-of-motion analysis for selected scalar shape descriptors. The analysis is based on a frequency of $2f_0$. Phase values in the plots are normalize to the full circle, i.e., $-180° \ldots 180°$ becomes $-0.5 \ldots 0.5$, and y_c is the phase reference. Figure 12 (a) and (b) show plots of the x_c and a_c phase features as a function of the amount of _shuffle_. Figure 12 (c) and (d) show the result obtained for varying _sexy walk_. There are similar plots for each of the 12 phase features. We show plots for only two features to reduce the amount of data for the reader.

All the phase features are consistent with the results of Little and Boyd [9], with the exception of x_c, x_{wc}, x_d, x_{uwc}, and x_{vwc} which are all out of phase by $180°$. We attribute this to the fact that Little and Boyd had pedestrians walk from right to left across the field of view, while our synthetic

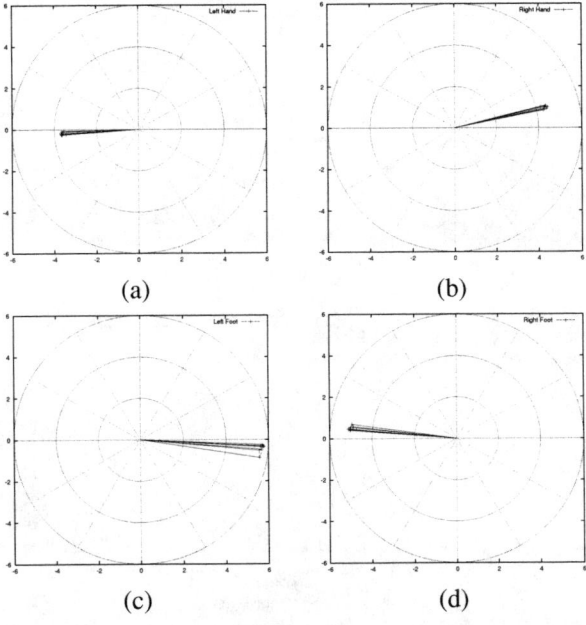

(a)　　　　　　(b)

(c)　　　　　　(d)

Figure 11. Phasor plots of joint movements for varying leg length (100 to 109%): (a) left hand, (b) right hand, (c) left foot, and (d) right foot. Radial lines are the phasors. The line connecting the ends of the phasors follows the change in leg length.

walker moves from left to right. This is sufficient to say that the synthetic data is at least a gross representation of "real" walkers.

Figures 12 (a) through (d) show ranges of feature values covering intervals of approximately 0.15 and 0.06 ($54°$ and $22°$) for x_c and 0.12 and 0.08 ($43°$ and $29°$) for a_c. These ranges are of the same order observed for the six subjects tested by Little and Boyd. Also, while Figures 12(a) and 12(d) suggest a linear relationship between *shuffle* and *sexy-walk*, and phase features, Figures 12(b) and 12(c) do not. In fact, we were not able to identify a pattern to the variation of gait style and phase feature. However, all features exhibited a feature range comparable to that measured by Little and Boyd.

Figure 12 (e) and (f) show the x_c and a_c phase features plotted as functions of the leg length of a pedestrian. The plots show ranges of feature values covering intervals of approximately 0.2 ($72°$) for x_c and 0.07 ($25°$) for a_c. Again, there is no discernible pattern in the relationship between phase feature and leg length, but the variation is large enough that recognition would be possible on this basis. We cannot explain this variation with the phase lock of the limb motion. Figure 11 indicates that the lock is not changing. Shape-of-motion appears to be sensitive to more than just phase lock. Here this may be due to the changing image area of the leg region and its effect on means computed over

the walking figure.

6. Discussion

Joint position data show clearly that phase lock changes with style of gait. This is not surprising since it is predicted by the psychophysical literature and is used as a basis for animation in computer graphics. In fact, the Poser software is likely changing the phase lock in order to simulate different gait styles. We do not observe a change in phase lock with the build of the walker. Again, this is likely due to the Poser software. In reality, we expect that a persons gait is affected by their build so that gait style and build are related, something not accounted for by Poser. To solve this, we expect that some sort of dynamic analysis would be required.

Our goal is to develop a model-free gait feature space that can be partitioned by subjective perception of the gait. What we can clearly see is that a change in style of gait requires a change in the phase lock of the gait, and that in turn causes variations in shape-of-motion phase features. Although we have a simple model that relates motion of points on limbs to the shape-of-motion features using phasors, the reality of gait perception is much more complex. We have already identified the non-linear rectification of the motion that doubles the frequency of limb motion, something that is impossible in a linear system. Occlusion presents another source of complexity. Even if the link from limb motion to shape-of-motion feature is complex, the variations in shape-of-motion features are definitely adequate for recognizing different types of gait.

Observed feature variations due to variations in kinematic structure are comparable to those for gait style. This presents a problem for a subjective feature space. We cannot tell whether or not a change in features is due to a change in kinematic structure (a different person) or a change in gait style (the same person walking differently). The shape-of-motion features take us part way to a subjectively partitioned feature space and are, in fact, adequate if only a single subject is considered.

Of course our results are based on a virtual person synthesized by some animation software. By doing this we forego the ability to form conclusions about real subjects. Our conclusions are only as good as the gait simulation done in Poser. However, we must live with them in order to conduct the sorts of tests shown here. It is not possible to set a dial on a human and obtain a smoothly varying amount of some subjective gait property, or to find a set of people that all walk the same way, and have identical kinematic structure with the exception of their legs. Simulation allows us to conduct an experiment that we could never do with live subjects.

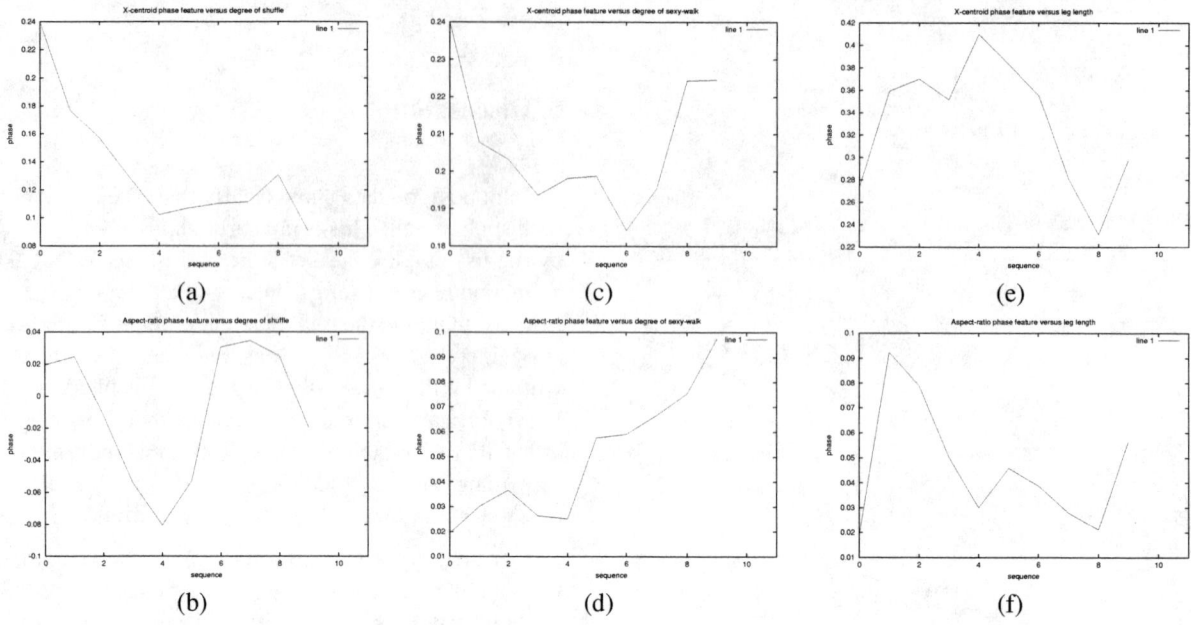

Figure 12. Normalized phase features, x_c and a_c as a function of gait property: *shuffle* (a,b), *sexy-walk* (c,d), and leg length (e,f). Sequence indices for *shuffle* or *sexy-walk* from 0 to 9 correspond to normal gait through 180%. Sequence indices for leg length correspond to 100% through 109% of normal leg length. a_c values (bottom row) are shifted by 0.5 ($180°$) to avoid phase wrap around.

7. Conclusions

We have shown that phase-sensitive shape-of-motion gait features are not only sensitive to kinematic structure, but are also sensitive to subjective variations in gait style. The observation can be explained by the changes in phase locking that are tied to any changes in a gait. Thus, shape-of-motion features provide a starting point in developing a gait feature space that is partitioned by subjective gait perception. Because the features are sensitive to kinematic structure, we can only subjectively partition the feature space where kinematic structure is fixed, i.e., for a single person. This suggests further development to eliminate the effects of varying kinematic structure.

References

[1] K. Amaya, A. Bruderlin, and T. Calvert. Emotion from motion. In *Graphics Interface 96*, 1996.

[2] I. Barrodale and R. E. Erickson. Algorithms for least-squares linear prediction and maximum entropy spectral analysis - part1: theory. *Geophysics*, 45(3):420–432, March 1980.

[3] B. I. Bertenthal and J. Pinto. Complementary processes in the perception and production of human movements. In L. B. Smith and E. Thelen, editors, *A Dynamic Systems Approach to Development: Applications*, pages 209–239. MIT Press, Cambridge, MA, 1993.

[4] J. E. Boyd and J. J. Little. Global versus structured interpretation of motion: moving light displays. In *IEEE Nonrigid and Articulated Motion Workshop, CVPR 97*, pages 18–25, 1997.

[5] H. Bulthoff, J. J. Little, and T. Poggio. A parallel algorithm for real-time computation of optical flow. *Nature*, 337:549–553, February 1989.

[6] R. Cutler and L. Davis. Robust periodic motion and motion symmetry detection. In *Computer Vision and Pattern Recognition 2000*, 2000.

[7] J. E. Cutting. A program to generate synthetic walkers as dynamic point-light displays. *Behavior Research Methods and Instrumentation*, 10(1):91–94, 1978.

[8] J. Laszlo, M. van de Panne, and E. Fiume. Limit cycle control and its application to the animation of balancing and walking. In *SIGGRAPH 96*, pages 155–162, 1996.

[9] J. J. Little and J. E. Boyd. Recognizing people by their gait: the shape of motion. *Videre*, 1(2):1–32, 1998.

[10] MetaCreations Corporation. *Poser 4 user guide*, 1999.

[11] M. P. Murray. Gait as a total pattern of movement. *American Journal of Physical Medicine*, 16(1):290–332, 1967.

[12] M. P. Murray, A. Bernard, and R. C. Kory. Walking patterns of normal men. *The Journal of Bone and Joint Surgery*, 46A(2):335–359, March 1964.

[13] R. J. Smith. *Circuits, Devices, and Systems*. John Wiley and Sons, Inc., New York, 3 edition, 1976.

[14] M. Unuma, K. Anjyo, and R. Takeuchi. Fourier principles for emotion-based human figure animation. In *SIGGRAPH 95*, pages 91–96, 1995.

A Computational Model for Motion Detection and Direction Discrimination in Humans

Yang Song[†] and Pietro Perona[††]

† California Institute of Technology, 136-93, Pasadena, CA 91125, USA

‡ Università di Padova, Italy

{yangs,perona}@vision.caltech.edu

Abstract

Seeing biological motion is very important for both humans and computers. Psychophysics experiments show that the ability of our visual system for biological motion detection and direction discrimination is different from that for simple translation [4]. But the existing quantitative models of motion perception can not explain these findings. We propose a computational model, which uses learning and statistical inference based on the joint probability density function (PDF) of the position and motion of the body, on stimuli similar to [4]. Our results are consistent with the psychophysics indicating that our model is consistent with human motion perception, accounting for both biological motion and pure translation.

1. Introduction

Perceiving the motion of the human body ('biological motion' in the literature of human vision) is a most important ability for the human visual system. Understanding how the brain perceives human motion and developing a computational model for it is an interesting and challenging problem for the fields of computer vision and human vision.

The abilities of the human visual systems for detection and direction discrimination, for both simple translation and biological motion, have been measured psychophysically [4]. In [4], a Johansson-like display [3] was used and it is found that the ability of the visual system to integrate biological motion over space and time is different from that of simple translation. Sensitivity to biological motion increases rapidly with the number of displayed joints, far more rapidly than for translation.

Many quantitative models of motion perception have been proposed, for example those in [5, 1, 9]. But they have been developed for translation, not for biological motion. No existing computational model can explain the difference between biological motion perception and translation perception as found in [4].

In [7, 6, 8], we proposed a perceptual model for detecting a moving human and for labeling its parts automatically. Rather than modeling the details of the mechanics of the human body, we choose to approach biological motion perception as the problem of recognizing a peculiar spatio-temporal pattern which may be learned perceptually. We observe the subject moving about in order to estimate a model of his/her stereotypical motions. This model, which we formulate as the joint probability density function (PDF) of the position and motion of the body, has a Markov-like structure.

The above model has demonstrated excellent and efficient performance on motion sequences with clutter and occlusion. It is therefore very interesting to compare the performance of it with that of the human visual system and examine if it can model how the human visual system behaves. In this paper we apply the probabilistic model to the tasks of detection and direction discrimination using stimuli similar to [4] and compare the results with the psychophysics results.

In section 2, the tasks and stimuli used to test the model are depicted. The probabilistic model is explained in section 3. Section 4 contains our simulation results, which are compared with psychophysics experiments in section 5.

2. Our stimuli

There are two kinds of tasks: one is to detect the presence of the target, and the other is to discriminate the direction of the target motion, both in the presence of dynamic random noise. The target is either a walker (biological motion) or simple translation. In the following, we first describe how the signals (targets) are generated, and then explain the tasks in more details.

11

2.1. Signals

Biological motion: We use the same program as in [4, 2] to generate the human walking sequence, where the motion of 13 dots represents the motion of the main joints of a person walking on a treadmill. Since we want to study how performance changes with the number of displayed joints, only a subset of the 13 joints appear in each frame. Each signal dot has a 'limited-lifetime' of two frames, then is 'reborn' at a randomly chosen joint, that is, we randomly select which joints to be displayed for each pair of frames, and during the whole sequences each joint has an equal chance to be represented.

Translation: Signal dots are generated in random positions over the area of the walker, with all moving at the same speed (set to match the average speed of the individual dots of biological motion). As in biological motion, the lifetime of each dot is also assumed to be two. Therefore, the positions of signal dots are generated randomly for the first frame, then those dots move to the second frame, and the positions are generated randomly again for the third frame, and so on.

2.2. Tasks

Detection. Detection is to decide which one of the two side by side displays contains the target: one consists of signal dots with certain amount of noise dots, and the other is a control display with the same dots density. Noise dots are generated independently for each frame using a uniform probability density. For biological motion, the control dots are derived from the walking algorithm [2] by randomizing the order of the frames presented. For translation, the control dots are generated independently for each frame.

Direction discrimination. Direction discrimination is to determine whether the target is moving rightwards or leftwards for a display known to contain the target. The display consists of signal dots superimposed with dynamic noise dots. The signal dots are generated as in section 2.1 for either biological motion or translation, and noise dots are generated randomly for each frame.

3. Computational Model

In the following subsections, the approaches of doing detection and direction discrimination from two frames are described. Based on the results from two frames, decisions upon multiple frames can be made handily as in [8].

For a pair of frames, positions and velocities of point features are taken as measurements, which are obtained from the local maxima of the Reichardt motion energy [5, 1, 9] between the two frames (see Appendix for our implementation).

3.1. Detection

Given two sets of measurements $\overline{X}_1$ and $\overline{X}_2$, detection is to decide which of the following two hypotheses is true:

Hypothesis 1 (O_1): $\overline{X}_1$ contains the target;

Hypothesis 2 (O_2): $\overline{X}_2$ contains the target.

Therefore, if $P(O_i|\overline{X}_i)$, $i = 1, 2$, is the posterior probability of the hypothesis O_i given $\overline{X}_i$, we need to compute the ratio

$$
\begin{aligned}
R(\overline{X}_1, \overline{X}_2) &= \frac{P(O_1|\overline{X}_1)}{P(O_2|\overline{X}_2)} \\
&= \frac{P(\overline{X}_1|O_1)P(O_1)/P(\overline{X}_1)}{P(\overline{X}_2|O_2)P(O_2)/P(\overline{X}_2)} \\
&= \frac{P(\overline{X}_1|O_1)}{P(\overline{X}_2|O_2)} \cdot \frac{P(O_1)}{P(O_2)} \cdot \frac{P(\overline{X}_2)}{P(\overline{X}_1)} \quad (1)
\end{aligned}
$$

where the second equal sign holds according to Bayes' law. If $R(\overline{X}_1, \overline{X}_2)$ is greater than 1, then $\overline{X}_1$ contains the target; otherwise the target is in $\overline{X}_2$. If the prior probabilities are assumed to be equal, the last two terms of the above equation are 1. As in [7, 8, 6], let $\overline{L}$ denote a possible labeling of $\overline{X}_1$ and assume $\mathcal{L}$ is all the possible labelings when $\overline{X}_1$ contains the target (O_1), then

$$
\begin{aligned}
P(\overline{X}_1|O_1) &= \sum_{\overline{L} \in \mathcal{L}} P(\overline{X}_1, \overline{L}|O_1) \\
&= \sum_{\overline{L} \in \mathcal{L}} P(\overline{X}_1|\overline{L}, O_1)P(\overline{L}|O_1) \quad (2)
\end{aligned}
$$

If we don't have any prior information about the labeling, then we can assume in the above equation, for any labeling $\overline{L}$, $P(\overline{L}|O_1) = 1/|\mathcal{L}|$, where $|\mathcal{L}|$ is the number of possible labelings. Let $\overline{X}_{fg}$ denote the foreground (target) measurements in $\overline{X}$, $\overline{X}_{bg}$ the measurements of background features, and $\overline{X}_{fg} \cup \overline{X}_{bg} = \overline{X}$. If foreground measurements and background measurements are independent,

$$
P(\overline{X}_1|\overline{L}, O_1) = P_{fg}(\overline{X}_{fg}) \cdot P_{bg}(\overline{X}_{bg}) \quad (3)
$$

If independent uniform background noise is assumed, $P_{bg}(\overline{X}_{bg})$ can be computed easily [8]. The computation of $P_{fg}(\overline{X}_{fg})$ will be described in section 3.3. $P(\overline{X}_2|O_2)$ can similarly be obtained.

3.2. Direction discrimination

For a given set of measurements $\overline{X}$, direction discrimination is to decide between the following two hypotheses:

Hypothesis 1 (H_1): rightward motion

Hypothesis 2 (H_2): leftward motion

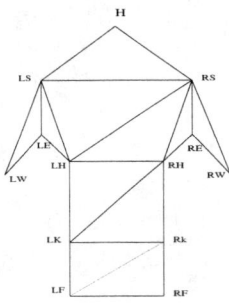

Figure 1. Decomposition of the human body into triangles [7]. 'L' and 'R' in label names indicate left and right. H:head, N:neck, S:shoulder, E:elbow, W:wrist, H:hip, K:knee and F:foot.

If $P(H_1|\overline{X}) > P(H_2|\overline{X})$, then the motion is rightwards, and vice versa.

$$
\begin{aligned}
\frac{P(H_1|\overline{X})}{P(H_2|\overline{X})} &= \frac{P(\overline{X}|H_1)P(H_1)/P(\overline{X})}{P(\overline{X}|H_2)P(H_2)/P(\overline{X})} \\
&= \frac{P(\overline{X}|H_1)}{P(\overline{X}|H_2)} \cdot \frac{P(H_1)}{P(H_2)}
\end{aligned}
\qquad (4)
$$

If we assume the prior probabilities are equal, i.e. $P(H_1) = P(H_2)$, then the decision rule becomes if $P(\overline{X}|H_1) > P(\overline{X}|H_2)$, the motion is rightwards, and vice versa. $P(\overline{X}|H_1)$ and $P(\overline{X}|H_2)$ can be computed in a similar way to equations (2) and (3).

3.3. Triangulated model for foreground probability

Biological motion [7, 8]. We first consider the case where all the body parts are present. By using the kinematic chain structure of human body, the whole body can be decomposed as in Figure 1. If the appropriate conditional independence (Markov property) is valid, then

$$
\begin{aligned}
&P_{fg}(\overline{X}_{fg}) \\
&= P_{LW,LE,LS}(X_{LW}|X_{LE},X_{LS})P_{LE,LS,LH}(X_{LE}|\dots) \\
&\quad \dots P_{RK,LF,RF}(X_{RK},X_{LF},X_{RF})
\end{aligned}
\qquad (5)
$$

where LW is the left wrist, RF is the right foot, etc; X_{LW} is the measurements (positions and velocities) of left wrist, X_{RF} is the measurements of right foot, etc. For our stimuli with some body parts missing in each frame, the foreground probability $P_{fg}(\overline{X}_{fg})$ is the marginalized version of equation (5) and can be computed as in [8]. Under this triangulated decomposition, the summation in equation (2) can be computed in polynomial time (on the order of N^4 where N is the number of observed features).

Translation. In case of no features missing, by the way of translation stimuli generated, the total number of signal dots in a translation display is the same as that of biological motion (13 here). To test the model, we assume that the

the joint foreground probability density function (PDF) for translation can also be decomposed into multiplications of joint (or conditional) PDF of triplets as in equation (5), i.e.,

$$
\begin{aligned}
&P_{fg}(\overline{X}_{fg}) \\
&= P_{A,B,C}(X_A|X_B,X_C)P_{B,C,D}(X_B|X_C,X_D) \\
&\quad \dots P_{K,L,M}(X_K,X_L,X_M)
\end{aligned}
\qquad (6)
$$

where $A, B, \dots, L, M$ are 13 labels, and $X_A, X_B, \dots, X_M$ are the corresponding measurements. Similarly to biological motion, when some features are missing (the number of signal dots is less than 13), the marginalized version ([8]) of the equation (6) is used to compute $P_{fg}(\overline{X}_{fg})$.

Though the probabilistic model structure and the computing method are the same for biological motion and translation, the model parameters (e.g. mean and covariance for Gaussian PDF) are different because the training sets are different (as will be explained in the next section).

4. Experiments

In our experiments, the probabilistic models are first learned, and then applied to the stimuli as described in section 2.

4.1. Training of the probabilistic models

Four kinds of foreground probabilistic model are learned: rightward and leftward motion for biological motion and translation respectively.

Biological motion. The training sequence is generated by the program in [4, 2]. For each pair of frames, positions and velocities are taken as measurements. Since the ground truth (labeled data) is needed for training, the velocities are obtained by subtracting the positions in two consecutive frames, not from Reichardt model. Independent uniform noise is added to both positions and velocities to match the quantization error introduced by the Reichardt detector which is used calculating velocities in the test data.

The training was done by estimating the joint (or conditional) probabilistic density functions (pdf) for all the triplets as described in section 3.3. As in [8, 6], we assumed all the pdfs were Gaussian, and the parameters for the Gaussian distribution were estimated from the training set.

Translation. For each frame, the positions of signal dots are generated randomly over the area, and the velocities are assigned to be the same for the dots in the same frame and generated from a uniform distribution over a certain range (identical to the range observed in biological motion, horizontal only and opposite for rightward and leftward model) across frames. Independent uniform noise is added to velocities to simulate the quantization error from Reichardt model.

13

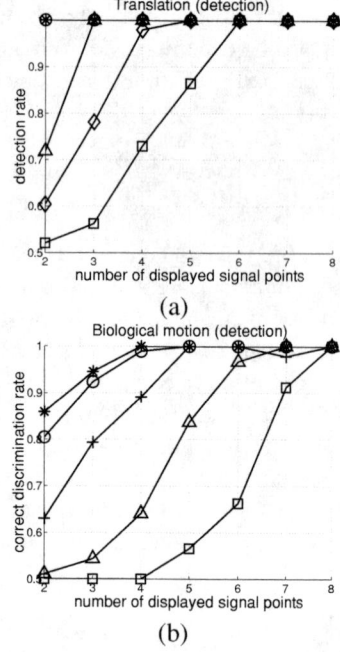

(a)

(b)

Figure 2. **Detection rate vs. number of displayed signal points (joints) under several noise levels. (a)** translation; **(b)** biological motion. The noise levels are: square 200 noise dots; diamond 150 noise dots (for translation only); triangle 100; plus 50; circle 30; star 10.

To use the decomposition model as in equation (6) for translation, labels are assigned randomly to signal dots for each pair of frames. The joint (or conditional) PDFs for all the triplets are assumed to be Gaussian.

4.2. Detection

The detection task, for both biological motion and translation, is performed on stimuli as in section 2. The size of the display is 170 by 310 pixels, a set-up very close to that in [4].

The algorithm described in section 3.1 is used. $\overline{X}_1$ and $\overline{X}_2$, which are positions and velocities for the image containing target and the control image, are obtained through Reichardt energy model so that $P(O_1|\overline{X}_1)$ and $P(O_2|\overline{X}_2)$ can be computed. In our simulations, we integrate over 5 pairs of frames to make decisions [8].

To compare with psychophysics results in [4], we study how sensitivity varies with the number of displayed signal dots (joints). As in [4], sensitivity is defined as the noise level (number of noise dots) at which 75% correct decisions are made. To find the sensitivity for a certain number of displayed signal dots, several noise levels were tried, and sensitivity was calculated by fitting a raised cumulative Gaussian curve (with asymptotes at 0.5 and 1) to the psychometric functions.

Figure 2 shows the detection rate vs. the number of dis-

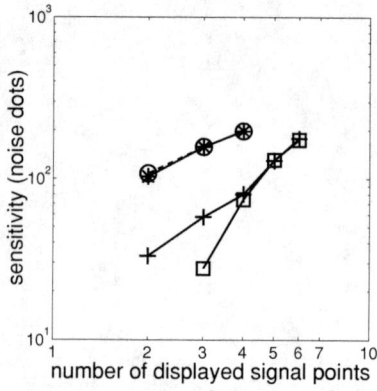

Figure 3. **Sensitivity vs. number of displayed joints.** Star (with dashed line): translation detection; circle (with dashed line): translation direction discrimination; plus (with solid line): biological motion detection; square (with solid line): biological motion direction discrimination.

played signal points (joints) under several noise levels, for (a) translation and (b) biological motion. For each condition (with a certain number of displayed signal points and a certain noise level), 360 frames (3 gait cycles) were used. The star (with dashed line) in Figure 3, derived from Figure 2(a), is the log-log sensitivity vs. number of displayed signal dots curve for translation detection, with a slope of 0.95 (calculated by linearly line fitting). The square (with solid line) in Figure 3, obtained from Figure 2(b), is the log-log curve for biological motion detection, with a slope of 1.53.

4.3. Direction discrimination

The direction discrimination task assumes that a moving target is in the scene and needs to decide the direction of the motion. For a given pair of images, positions and velocities are obtained using Reichardt model as measurements ($\overline{X}$), and plugged into $P(\overline{X}|H_1)$ and $P(\overline{X}|H_2)$ as described in section 3.2. As in detection, decisions are then made upon integration over 5 pairs of frames. Sensitivities are calculated in the same way as in section 4.2.

The curves in Figure 4 are the correct direction discrimination rate vs. the number of displayed signal points (joints) under several noise levels, for (a) translation and (b) biological motion. The circle (with dashed line) in Figure 3, derived from Figure 4(a), is the log-log curve of sensitivity vs. number of displayed signal dots for translation direction discrimination, with a slope of 0.88. The plus (with solid line) in Figure 3, obtained from Figure 4(b), is the log-log curve for biological direction discrimination, with a slope of 2.71.

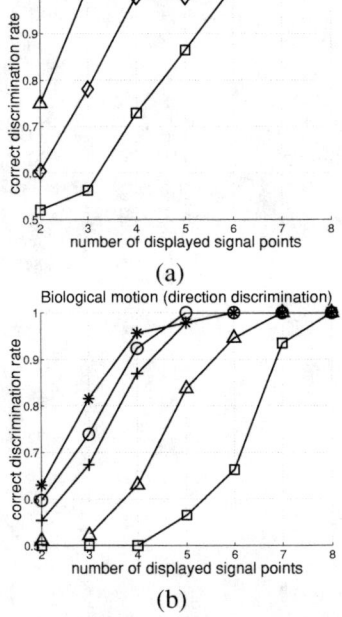

(a)

(b)

Figure 4. **Correct direction discrimination rate vs. number of displayed signal points (joints) under several noise levels. (a)** translation; **(b)** biological motion. The noise levels are: square 200 noise dots; diamond 150 noise dots (for translation only); triangle 100; plus 50; circle 30; star 10.

5. Comparison with psychophysics

From Figure 3, the log-log curves for perceiving biological motion are steeper than those for translation, that is, the performance of biological motion perception changes more rapidly with the number of displayed signal points than that of translation, which is consistent with the psychophysics results in [4]. This indicates that the number of displayed signal points is more crucial to biological motion perception, which implies that to perceive biological motion, more concerted movement (collaboration) among signal dots is needed. The steepest curve is the one of biological motion direction discrimination, which conforms with our intuition (also the results of psychophysics) that the direction perception of human motion is the most demanding task and needs the signal dots to be the most concerted (collaborative).

The above observation can be explained intuitively by our probabilistic model. For biological motion, the relative positions and velocities are correlated in the PDF of one triplet. But for translation, relative positions and velocities are independent, velocities of different parts are highly correlated and with small variance, and positions are almost independent and with large variance. Therefore, for biological motion, if only two signal points with big relative distance (not in the same triangle) are observed, it is very unlikely for our probabilistic model to take them as a human configuration. For translation, if two signal points are

observed, then regardless of their relative position, they can give a higher likelihood being translation as long as they have similar velocities. So in some sense, when the number of signal dots is small, the dots of translation are more 'informative' than those of biological motion.

Our experimental set-up is very similar to that of [4], but different from theirs in the temporal integration part. In their paper, they used the 'limited-lifetime' technique and integration over 1200ms (40 frames). In our temporal integration, we assume independence among pairs of frames, and only integrate over 5 pairs. We believe that experiments with the same condition as them would be qualitatively similar.

6. Conclusions

The consistency between our results and the psychophysics both of biological motion and translational motion perception suggests that our model could be a good computational model for human motion perception.

Our probabilistic model indicates that the visual system may gain the ability of perceiving biological motion and translation through learning. The mechanisms for perceiving biological motion and translation could be the same, but are tuned to different model parameters. When biological motion is perceived, it may not be viewed as a whole, but some closer (or more correlated) body parts may be grouped together first.

Our model could predict the performance of the human visual system on any complex motion pattern. Detailed comparison of such predictions with the psychophysics would allow further refinements of the model.

Appendix: implementation of Reichardt-type feature velocities between two frames [5, 1, 9]

This appendix describes our implementation of getting point feature velocities between two frames using a Reichardt-type model. A image sequence can be represented as a function $I(x, y, t)$, where x and y are spatial coordinates in horizontal and vertical directions respectively, and t is the time coordinate. We compute the velocities between two frames $I(x, y, t)$ and $I(x, y, t+1)$ (for simplicity the time interval is assumed to be 1) in three steps:

(1) Spatial filtering is first applied to both images. Let K(x,y) denote the filter (we use the same filter for both images), then,

$$f_1(x, y) \stackrel{\text{def}}{=} I(x, y, t) * K(x, y)$$
$$f_2(x, y) \stackrel{\text{def}}{=} I(x, y, t+1) * K(x, y)$$

where $*$ means convolution, and $f_1(x, y)$ and $f_2(x, y)$ are the two images after spatial filtering.

15

(2) Get motion energy under different velocities. Let vx and vy be respectively the horizontal and vertical velocities between the two frames, then $E(x, y, vx, vy)$, which is the motion energy for velocity (vx, vy) at location (x, y), is computed as

$$E(x, y, vx, vy) = f_1(x, y) \cdot f_2(x + vx, y + vy)$$

(3) The local maxima of $E(x, y, vx, vy)$ are taken as the feature velocities between the two frames, that is, velocity (vx_i, vy_i) can be perceived at location (x_i, y_i) if

$$\frac{\partial E}{\partial x}|_{(x_i, y_i, vx_i, vy_i)} = 0, \qquad \frac{\partial E}{\partial y}|_{(x_i, y_i, vx_i, vy_i)} = 0,$$

$$\frac{\partial E}{\partial vx}|_{(x_i, y_i, vx_i, vy_i)} = 0, \quad \text{and} \quad \frac{\partial E}{\partial vy}|_{(x_i, y_i, vx_i, vy_i)} = 0$$

(x_i, y_i, vx_i, vy_i)'s are positions and velocities of point features between the two frames.

Note that in our implementation, x, y, vx and vy are all discretized, therefore, the resolution of the features (x_i, y_i, vx_i, vy_i) depends on the quantization scale. Also, energy E is only computed for a certain range of (vx, vy), which limits the range of a feature velocity (vx_i, vy_i) can be in.

Acknowledgments

Funded by the NSF Engineering Research Center for Neuromorphic Systems Engineering (CNSE) at Caltech (NSF9402726), and by an NSF National Young Investigator Award to PP (NSF9457618). We thank Peter Neri for providing the code of generating the human walking sequences [2].

References

[1] E. Adelson and J. Bergen. Spatiotemporal energy models for the perception of motion. *J. Opt. Soc. Am. A*, 2:284–299, 1985.

[2] J. Cutting. A program to generate synthetic walkers as dynamic point-light displays. *Behav. Res. Methods Instrument*, 10:91–94, 1978.

[3] G. Johansson. Visual perception of biological motion and a model for its analysis. *Perception and Psychophysics*, 14:201–211, 1973.

[4] P. Neri, M. Morrone, and D. Burr. Seeing biological motion. *Nature*, 395:894–896, 1998.

[5] W. Reichardt. Autocorrelation, a principle for the evaluation of sensory information by the central nervous system. In *Sensory Communication, W.A. Rosenblith, ed.* Wiley, New York, 1961.

[6] Y. Song, X. Feng, and P. Perona. Towards detection of human motion. In *Proc. IEEE CVPR*, volume 1, pages 810–817, June 2000.

[7] Y. Song, L. Goncalves, E. D. Bernardo, and P. Perona. Monocular perception of biological motion - dection and labeling. In *International Conference on Computer Vision*, pages 805–812, Sept 1999.

[8] Y. Song, L. Goncalves, and P. Perona. Monocular perception of biological motion - clutter and partial occlusion. In *Proc. ECCV*, volume 2, pages 719–733, June/July 2000.

[9] J. van Santen and G. Sperling. Elaborated reichardt detectors. *J. Opt. Soc. Am. A*, 2:300–321, 1985.

Session 2

Direction & Tracking II

Specialized Mappings and the Estimation of Human Body Pose from a Single Image

Rómer Rosales and Stan Sclaroff
Boston University, Computer Science Department
111 Cummington St., Boston, MA 02215
email:{rrosales,sclaroff}@bu.edu

Abstract

We present an approach for recovering articulated body pose from single monocular images using the Specialized Mappings Architecture (SMA), a non-linear supervised learning architecture. SMA's consist of several specialized forward (input to output space) mapping functions and a feedback matching function, estimated automatically from data. Each of these forward functions maps certain areas (possibly disconnected) of the input space onto the output space. A probabilistic model for the architecture is first formalized along with a mechanism for learning its parameters. The learning problem is approached using a maximum likelihood estimation framework; we present Expectation Maximization (EM) algorithms for several different choices of the likelihood function. The performance of the presented solutions under these different likelihood functions is compared in the task of estimating human body posture from low level visual features obtained from a single image, showing promising results.

Figure 1. The data used for training is formed by 2D marker positions and their corresponding image visual features. Here we show some frames from the same sequence viewed from two given camera orientations (a) **0** rads, (b) **6Π/32** rads. Training is done sampling the set of all possible orientations (here 32) from the same distance and height.

1 Introduction and Related Work

Estimating articulated body pose from low-level visual features is an important yet difficult problem in computer vision and machine learning. To date, there has been extensive research in the development of algorithms for human motion tracking [7, 21, 19, 4, 13, 9, 23, 17] and recognition [5], human pose estimation from a single image [1, 20], and machine learning approaches [3, 12, 22, 20]. Being able to infer detailed body pose, would open the doors to the development of a great number of applications for human-computer interfaces, video coding, visual surveillance, human motion recognition, ergonomics, and video indexing/retrieval, etc.

In their everyday life, humans can easily estimate body part location and structure from relatively low-resolution images of the projected 3D world (*e.g.*,watching a video). Unfortunately, this problem is inherently difficult for a computer. Finding the mapping between low-level image features and body configurations is highly complex and ambiguous. The difficulty stems from the number of degrees of freedom in the human body, the complex underlying probability distribution, ambiguities in the projection of human

motion onto the image plane, self-occlusion, insufficient temporal or spatial resolution, etc.

In this paper we attack the problem of articulated body pose estimation within the framework of non-linear supervised learning. In particular, we use a novel machine learning architecture, the Specialized Mappings Architecture (SMA). This SMA's fundamental components are a set of specialized mapping functions, and a single feedback matching function. All of these functions are estimated directly from data, in our case: examples of body poses (output) and their corresponding visual features (input).

SMA's are related to machine learning models [14, 11, 8, 20] that use the principle of divide-and-conquer to reduce the complexity of the learning problem by splitting it into several simpler ones. In our case, each of these hopefully simpler problems is attacked using different specialized functions that act as the simpler problem solvers.

In general these algorithms try to fit surfaces to the observed data by (1) splitting the input space into several regions, and (2) approximating simpler functions to fit the input-output relationship inside these regions. Sometimes these functions can be constants, and the regions may be recursively subdivided creating a hierarchy of functions. Convergence has been reported to be generally faster than gradient-based neural network optimization algorithms [14].

The divide process may create a new problem: how to optimally partition the problem such that we obtain several sub-problems that can be solved using the specific solver capabilities (*i.e.*,form of mapping functions). In this sense we can consider [20] as a simplification of our approach, where the splitting is done at once without considering neither the power or characteristics of the mapping functions nor input-output relationship in the training set. This gives rise to two independent optimization problems in which input regions are formed and a mapping function estimated for each region, causing sub-optimality. In this paper we generalize these underlying ideas and present a probabilistic interpretation along with a estimation framework that simultaneously optimizes for both problems. Moreover, we provide a formal justification of the seemingly ad-hoc method described in [20].

In the work of [8], *hard* splits of the data were used, *i.e.*,the parameters in one region only depend on the data falling in that region. In [14], some of the drawbacks of the hard-split approach were pointed out (*e.g.*,increase in the variance of the estimator), and an architecture that uses *soft* splits of the data, the Hierarchical Mixture of Experts, was described. In this architecture, as in [11], at each level of the tree, a gating network is used to control the influence (weight) of the expert units (mapping functions) to model the data. However, in [11] arbitrary subsets of the experts units can be chosen. Unlike these architectures, in SMA's the mapping selection is done using a feedback matching process, currently in a winner-take-all fashion, but *soft* splitting is done during training.

Previous learning based approaches for estimating human body pose include [12], where a statistical approach was taken for reconstructing the three-dimensional motions of a human figure. It consisted of building a Gaussian probability model for short human motion sequences. This method assumes that 2D tracking of joints in the image is given. Unlike this method, we do not assume tracking can be performed (*e.g.*,we do not assume that a body model can be matched to images from frame to frame). There are many known disadvantages and limitations in performing visual tracking [20]: manual initialization, poor long-term stability, necessary iterative solutions during reconstruction, high dependence of algorithms and characterisitics of the articulated model.

In [3], the manifold of human body configurations was modeled via a hidden Markov model and learned via entropy minimization. In [22] dynamic programming is used to calculate the best global labelling of the joint probability density function of the position and velocity of body features; it was assumed that it is possible to track these features for pairs of frames.

Unlike these previous learning based methods, our method does not attempt to model the dynamical system; instead, it relies only on instantaneous configurations. Even though this ignores information (*i.e.*,motion components) that can be useful for constraining the reconstruction process, it provides invariance with respect to speed (*i.e.*,sampling differences) and direction in which motions are performed. Furthermore, fewer training sequences are needed in learning a model. In our approach, a feedback matching step is used, which transforms the reconstructed configuration back to the visual cue space to choose among the set of reconstruction hypotheses. Finally, no tracking is assumed.

2 Specialized Mappings and Learning

In this paper, SMA's are described to approach the problem of supervised learning. Define the set of output-input observations pairs $\mathcal{Z} = \{(\psi_i, v_i)\}$, with $\psi_i \in \Psi$ and $v_i \in \Upsilon$. Let us call the output and input vectors the target and cue vectors and consider them as elements of $\Re^t$ and $\Re^c$ respectively.

Let us assume that there is a functional relation between cue and target vectors that we call $\phi^\star : \Re^c \to \Re^t$, such that $\psi_i \approx \phi^\star(v_i)$, define this to be the forward mapping. The problem is to approximate this function $\phi^\star$.

In theory this problem can be formulated by finding $\phi^* = \arg\min_\phi \sum_{i=1}^n \rho(\phi(v_i) - \psi_i)$ where n is the cardinality of Ψ or Υ [2, 10, 15], and ρ is an error function. The problem of function approximation from sparse data is known to be ill-posed if no further constraints are added [2, 10] (*e.g.*,on the functional form or architecture of ϕ).

In this paper, we attack nonlinear supervised learning problems using an architecture that generates a series of m functions ϕ_k in which each of these functions is specialized to map only certain inputs, for example a region of the input space. However, the domain of ϕ_k can be more general than just a connected region in the input space. We propose to determine these regions and functions simultaneously.

In contrast with [14, 11] we do not have a mixture of expert functions weighted by gating networks when generating an output, in SMA's, an input is only mapped by a given function. For this, assume there is another functional relation such that $v_i \approx \zeta(\psi_i)$ (*i.e.*,an inverse mapping), which can be known, or learned. Given this, SMA's involve a feedback matching process to choose among the series of hypotheses given by each specialized function.

2.1 Probabilistic Model

In order to give a probabilistic interpretation to the architecture, let's define some notation first. Let the training sets of output-input observations be $\Psi = \{\psi_1, \psi_2, ..., \psi_n\}$, and $\Upsilon = \{v_1, v_2, ..., v_n\}$ respectively. We will use $\mathbf{z}_i = (\psi_i, v_i)$ to define the given output-input training pair, and $\mathcal{Z} = \{\mathbf{z}_1...\mathbf{z}_n\}$ as our observed training set. In general the vector $\mathbf{z}$ is defined to be composed of two parts, one denoted ψ and another denoted v associated with the output and input space respectively.

Define the unobserved random variables $\mathbf{Y}_i$ with $i = \{1..n\}$. In our model these variables have domain the discrete set $\mathcal{C} = \{1..m\}$ of labels for the specialized functions, and can be thought as the function number used to map data point i, therefore m is the number of specialized functions in the model.

Our model uses parameters $\theta = (\theta_1, \theta_2, ...\theta_m, \lambda)$, where θ_i represents the parameters of the mapping function i. The vector $\lambda = (\lambda_1, \lambda_2, ..., \lambda_m)$, where λ_k represent $P(y_i =$

$k|\theta)$, the prior probability that mapping function with label i will be used to map an unknown point.

As an example, $P(y_i|z_i,\theta)$ represents the probability that function number y_i generated data point number i (given our model parameters).

Using Bayes' rule and assuming independence among observations and an uniform prior $p(\theta)$ we have the joint probability of our architecture:

$$P(\mathcal{Z},\mathbf{y},\theta) = P(\mathcal{Z}|\mathbf{y},\theta)P(\mathbf{y}|\theta) = \prod_i P(\mathbf{z}_i|y_i,\theta)P(y_i|\theta) \tag{1}$$

A key question in instantiating the architecture is: What is $P(\mathbf{z}|y,\theta)$? (the probability that point $\mathbf{z}$ was generated using the mapping function y assuming a certain value for its parameters). In this paper we analyze three possible cases:

1. A Gaussian joint distribution of input-output vectors:

$$P(\mathbf{z}|y,\theta) = P(\psi,v|y,\theta) = \mathcal{N}((\psi,v);\mu_y,\Sigma_y) \tag{2}$$

2. A Gaussian distribution with mean defined by the error incurred in using the possibly non-linear function ϕ_y as a mapping function, and a fixed, given variance Σ_y.

$$P(\mathbf{z}|y,\theta) = \mathcal{N}(\psi;\phi_y(v,\theta),\Sigma_y) \tag{3}$$

3. A comparison of distance measures among all functions, it generates a competition among functions to represent the data points, for example:

$$P(\mathbf{z}_i|y_i,\theta) = \frac{e^{-\rho(\psi_i - \phi_{y_i}(v_i,\theta))}}{e^{-\sum_{y_i}\rho(\psi_i-\phi_{y_i}(v_i,\theta))}}, \tag{4}$$

where ρ is a given error norm, and ϕ_j is the j-th mapping function. This can be written more generally as:

$$P(\mathbf{z}_i|y_i,\theta) = \frac{e^{-\chi_{y_i}(\mathbf{z}_i,\theta)}}{e^{-\sum_{y_i}\chi_{y_i}(\mathbf{z}_i,\theta)}} \tag{5}$$

3 EM algorithms for Learning the Parameters of the Model

The probabilistic parameter estimation problem is approached under the Expectation Maximization (EM) algorithm framework [6] using the notation followed by [16]. The E-step consists of finding $\tilde{P}(\mathbf{y}) = P(\mathbf{y}|\mathcal{Z},\theta)$. It can be shown that this reduces to:

$$\tilde{P}(\mathbf{y}) = \prod_i \frac{\lambda_{y_i}P(\mathbf{z}_i|y_i,\theta)}{\sum_{k\in\mathcal{C}}\lambda_k P(\mathbf{z}_i|y_i=k,\theta)} = \prod_i \tilde{P}^{(t)}(y_i) \tag{6}$$

The M-step consists of finding $\theta^{(t)} = \arg\max_\theta E_{\tilde{P}^{(t)}}[\log P(\mathbf{y},\mathcal{Z}|\theta)]$. In our case we can show that this is equivalent to:

$$\theta^{(t)} = \arg\max_\theta \sum_i \sum_{y_i\in\mathcal{C}} \tilde{P}^{(t)}(y_i)[\log P(\mathbf{z}_i|y_i,\theta)+\log P(y_i|\theta)]. \tag{7}$$

It is important to mention that this is valid if $P(\mathbf{z}_i|\theta)$ depends on y_i and not on y_j, for any $j \neq i$. Note that for the distributions discussed above, this is true. We present solutions for the cases described above. Due to space constraints, only final equations are shown. **In case (1) we have:**

$$P(\mathbf{z}|y,\theta) = \mathcal{N}(\mu_y,\Sigma_y) = \mathcal{N}(\begin{bmatrix}\mu_v\\\mu_\psi\end{bmatrix},\begin{bmatrix}\Sigma_{vv}\ \Sigma_{v\psi}\\\Sigma_{v\psi}^\top\ \Sigma_{\psi,\psi}\end{bmatrix})_y \tag{8}$$

In this case, we can show that the SMA architecture parameter learning problem is neatly reduced to mixture of Gaussian estimation, for which it is straightforward to estimate θ using EM. Moreover, the ML estimate of the conditional distribution (the conditional distribution is of major importance because our problem consist in esimating ψ from observing v) $P(\psi|v,y,\theta)$ is also Gaussian, given by:

$$P(\psi|v,y,\theta) = \mathcal{N}(\mu_\psi+\Sigma_{v\psi}^\top\Sigma_{vv}^{-1}(v-\mu_v),\Sigma_{\psi\psi}-\Sigma_{v\psi}^\top\Sigma_{vv}^{-1}\Sigma_{v\psi})_y \tag{9}$$

Therefore in case (1), each specialized function ϕ_k is just the mean of the conditional distribution (conditioned on the observation v_i and the function index);

$$\phi_k(v,\theta) = \mu_\psi + \Sigma_{v\psi}^\top\Sigma_{vv}^{-1}(v - \mu_v), \tag{10}$$

moreover we have an expression for the confidence on this estimate given by the variance above. Thus, the set of functions ϕ_k are linear in the input vector.

In case (2) we have:

$$\frac{\partial E}{\partial \lambda_k} = \sum_i \tilde{P}^{(t)}(y_i = k)\frac{\partial}{\partial\lambda_k}\log P(y_i = k|\theta) \tag{11}$$

$$\frac{\partial E}{\partial \theta_k} = \sum_i \tilde{P}_i^{(t)}(y_i = k)$$
$$[(\frac{\partial}{\partial\theta_k}\phi_k(v_i,\theta_k))^\top \Sigma_k^{-1}(\psi_i - \phi_k(v_i,\theta_k))] \tag{12}$$

where E is the cost function found in Eq. 7. This gives the following update rule for λ_k (where Lagrange multipliers were used to incorporate the constraint $\sum_k \lambda_k = 1$).

$$\lambda_k = \frac{1}{n}\sum_i P(y_i = j|\mathbf{z}_i,\theta) \tag{13}$$

The update of θ_k depends on the form of ϕ_k.

This case is of particular importance in justifying the approach presented in [20] from a probabilistic perspective. In [20] output data (from $\mathbf{\Psi}$) is clustered using a mixture of Gaussians models, and then for each cluster a multi-layer perceptron is used to estimate the mapping from input to output space. Let us consider the SMA obtained by choosing ϕ_k to be a multi-layer perceptron neural network. First note that the bracketed term in Eq. 12 is equivalent to back-propagation (assuming $\Sigma_k = \mathbf{I}$).

Using a winner-take-all variant to update the gradient found in Eq. 12, we have:

$$\frac{\partial E}{\partial \theta_k} = \sum_{i\in W_k}[(\frac{\partial}{\partial\theta_k}\phi_k(v_i,\theta_k))^\top\Sigma_k^{-1}(\psi_i - \phi_k(v_i,\theta_k))] \tag{14}$$

21

with $W_k = \{i | \arg\max_j \tilde{P}^{(t)}(y_i = j) = k\}$ (i.e.,use a hard assignment of the data points to optimize each of the functions, according to the posterior probability $\tilde{P}^{(t)}$). Therefore we have that each of the specialized functions is trained using backpropagation with a subset of the training sets (moreover these subsets are disjoint)

Note that the maximization process that finds the sets W_k can also be stated as

$$\arg\max_j P(z_i | y_i = j, \theta) P(y_i = j | \theta) \qquad (15)$$

The approach in [20] can then be explained within the framework of SMA's presented here by (1) performing the E-step (i.e.,computing $\tilde{P}^{(t)}(y_i)$) once and therefore fixing $\tilde{P}^{(t)}(y_i)$ throughout the whole optimization process, (2) using a winner-take-all variant for the M-step. Finally, (3) the choice of a Gaussian cost function for clustering (done in the E-step) is justified by choosing $P(z_i | \theta)$ to be a Gaussian mixture, as suggested by Eq. 15. Let us call this special version of case (2), case (2a).

In case (3) we have: Taking derivatives in Eq. 7 with respect to λ_k we obtain Eq. 13 as the update rule for λ_k. Taking derivatives in Eq. 7 with respect to θ, we obtain:

$$\frac{\partial E}{\partial \theta_k} = \sum_i \{ \frac{\partial}{\partial \theta_k} \chi_k(z_i, \theta_k)$$

$$[P(z_i | y_i = k, \theta_k) - \tilde{P}_i^{(t)}(y_i = k)] \qquad (16)$$

Note that, in keeping the formulation general, we have not defined the form of the specialized functions ϕ_k in Eqs. 12 and 16. In both cases whether or not we can find a closed form solution for the update of θ_k depends on the form of ϕ_k. For example if ϕ_k is a non-linear function, it is likely that we may have to use iterative optimization to find $\theta_k^{(t)}$. In the case where ϕ_k yields a quadratic form for χ_k then a closed form update exists. Note also that the bracketed term in Eq. 16 is the difference between prior and posterior distributions (which gives an intuition on what the goal of the process is), and only affects the *importance* or weight of the contribution of each data point. The results from case (3) will be evaluated experimentally in further work.

4 Feedback Matching

When generating an output $\hat{y}$ given an input $\mathbf{x}$, we have a series of output hypotheses $\hat{Y}$ obtained using $\hat{y}_k = \phi_k(\mathbf{x})$, with $k \in \mathcal{C}$. Given the set $\hat{Y}$, we define the most accurate hypothesis to be that one that minimizes a function $F(\zeta(\hat{y}_j), \mathbf{x}, \mathcal{Z})$, over j for example:

$$i = \arg\min_j (\zeta(\hat{y}_j) - \mathbf{x})^\top \Sigma_\Upsilon^{-1} (\zeta(\hat{y}_j) - \mathbf{x}), \qquad (17)$$

where Σ_Υ is the covariance matrix of the elements in the set Υ and i is the assigned label. It is important to notice that the feedback matching could be used actively during learning instead of using it only during inference to choose among the set of hypotheses. The form of the cost function could vary, here (in Eq. 17)we have assumed that the data from Υ is Gaussian distributed. This is explained more thoroughly in [20].

5 Experiments

Cases (1), (2) and (2a) of the described SMA formulation were tested. The experimental setup is the same as that used in [20]. We used a computer graphics based feedback function ζ [20], and Eq. 17 as feedback matching cost function.

The training data consisted of twelve sequences obtained through 3D motion capture. As stated previously, training data consists of set of example input-output pairs, (ψ_i, v_i). The output consisted of 11 2D marker positions (projected to the image plane using a perspective model) but linearly encoded by eight real values using Principal Component Analysis (PCA). The input consisted of seven real-valued Hu moments computed on synthetically generated silhouettes of the articulated figure. Input-output pairs were generated using computer graphics by sampling the equator of the view-sphere to render 32 views [20].

We generated approximately 60,000 data vectors for training (corresponding to 32 views) and 9,984 for testing (also containing samples, equally distributed, from 32 views). The only free parameters in this test, related to the given SMA's, were (a) the number of specialized functions used: 15, 5, 15 for cases (1) (2) and (2a) respectively and (b) for case (2) and (2a) we chose ϕ_k to be multi-layer perceptrons with 16 hidden neurons. Note that several model selection approaches could be used instead to choose the number of parameters of the architecture (e.g.,Minimum Description Length [18]).

Fig. 2 shows the body pose estimates obtained in several single images coming from two different sequences at specific orientations (due to space limitations case (2) is not included, in this case its performance is comparable with the rest). The agreement between pose estimates and ground-truth is easy to perceive for all sequences. Note that for self-occluding configurations, pose estimation is harder, but still the estimate is close to ground-truth. No human intervention nor pose initialization is required.

Using the training and testing data described above, we measured the average marker error for both models (as the distance between reconstructed and ground-truth projected marker position). With respect to the height of the body, the mean and variance marker error were: (1) 2.82% and 0.09%, (2) 2.73% and 0.02%, (2a) 2.34% and 0.04% respectively. Note the number of parameters in each model: (1) 3600 (2) 1205 (2a) 3615 . In case (1), the training was considerably faster because of the extra processing time necessary in (2) and (2a) for training each neural network once the clustering or weights per sample is decided. The smaller variance obtained in case (2) (in general a desirable behavior) is probably due to the *soft* splits of the data used by the learning algorithm. Inference required approximately the same computational time per specialized function in each case.

Fig. 3 shows the average marker error and variance per body orientation. Note that in all cases the error is bigger for orientations closer to $\pi/2$ and $3\pi/2$ radians. This intuitively agrees with the notion that at those angles (side-view), there is less visibility of the body parts.

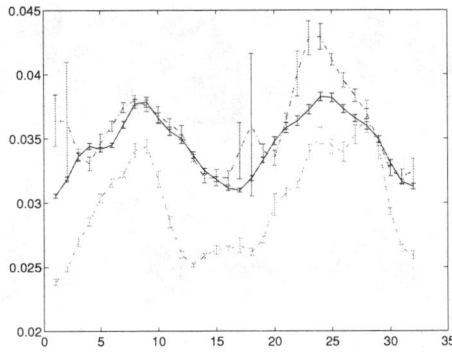

Figure 3. Mean marker error and variance for cases (1) (top-broken), (2) (middle-continuous) and (2a) (bottom-broken) per view angle, sampled every $2\pi/32$ radians.

5.1 Experiments using Real Visual Cues

For the next example, in Fig. 4 we test the system against real segmented visual data, obtained from observing a human subject. Reconstruction for several relatively complex action sequences is shown for both models. Note that even though the characteristics of the segmented body differ from the ones used for training, good performance is achieved. Most frames are visually close to what can be thought as the right pose reconstruction. Body orientation is also correct.

The following variables are believed to account the most in performance: 1.) likelihood distribution choice 2.) enough data to account for observed configurations 3.) number of approximating functions with specialized domains, 4.) differences in body characteristics used for training/testing, and 5.) discriminative power of the chosen image features (Hu moments reduce the image interpretation to a seven-dimensional vector).

6 Conclusion

We have proposed the use of a non-linear supervised learning framework, Specialized Mappings Architecture (SMA), for estimating human body pose from single images. A learning algorithm was developed for this architecture using the framework of ML estimation, latent variable models and Expectation Maximization. The implemented algorithm for inference runs in linear time $O(M)$ with respect to the number of specialized functions M.

The incorporation of the feedback step actively during learning is an important possibility provided by SMA's and currently being considered. Note that so far the feedback matching is used for inference only (for choosing among the set of hypotheses). Feedback could also be used for determining the distribution or importance of each training sample with respect to each of the mapping functions.

In experiments, a SMA learns how to map low-level visual features to a higher level representation like a set of joint positions of the body. Human pose reconstruction from a single image is a particularly difficult problem because this mapping is highly ambiguous and complex. We have obtained excellent results even using a very simple set of image features, such as image moments. Choosing the best subset of image features from a given set is by itself a complex problem, and a topic of on-going research. This is a very important step considering that low-level visual features are relatively easily obtained using current vision techniques.

References

[1] C. Barron and I. Kakadiaris. Estimating anthropometry and pose from a single image. In *CVPR*, 2000.

[2] M. Bertero, T. Poggio, and V. Torre. Ill-posed problems in early vision. *Proc. of the IEEE*, (76) 869-889, 1988.

[3] M. Brand. Shadow puppetry. In *ICCV*, 1999.

[4] C. Bregler. Tracking people with twists and exponential maps. In *CVPR98*, 1998.

[5] J. Davis and A. F. Bobick. The representation and recognition of human movement using temporal templates. In *CVPR*, 1997.

[6] A. Dempster, N. Laird, and D. Rubin. Maximum likelihood estimation from incomplete data. *Journal of the Royal Statistical Society (B)*, 39(1), 1977.

[7] J. Deutscher, A. Blake, and I. Reid. Articulated body motion capture by annealed particle filtering. In *CVPR*, 2000.

[8] J. Friedman. Multivatiate adaptive regression splines. *The Annals of Statistics*, 19,1-141, 1991.

[9] D. Gavrila and L. Davis. Tracking of humans in action: a 3-d model-based approac. In *Proc. ARPA Image Understanding Workshop, Palm Springs*, 1996.

[10] F. Girosi, M. Jones, and T. Poggio. Regularization theory and neural network architectures. *Neural Computation*, (7) 219-269, 1995.

[11] G. Hinton, B. Sallans, and Z. Ghahramani. A hierarchical community of experts. *Learning in Graphical Models, M. Jordan (editor)*, 1998.

[12] N. Howe, M. Leventon, and B. Freeman. Bayesian reconstruction of 3d human motion from single-camera video. In *NIPS*, 1999.

[13] L. D. I. Haritaouglu, D. Harwood. Ghost: A human body part labeling system using silhouettes. In *Intl. Conf. Pattern Recognition*, 1998.

[14] M. I. Jordan and R. A. Jacobs. Hierarchical mixtures of experts and the em algorithm. *Neural Computation*, 6, 181-214, 1994.

[15] N. Kolmogorov and S. Fomine. *Elements of the Theory of Functions and Functional Analysis*. Dover, 1975.

[16] R. Neal and G. Hinton. A view of the em algorithm that justifies incremental, sparse, and other variants. *Learning in Graphical Models, M. Jordan (editor)*, 1998.

[17] J. M. Regh and T. Kanade. Model-based tracking of self-occluding articulated objects. In *ICCV*, 1995.

[18] J. Rissanen. Stochastic complexity and modeling. *Annals of Statistics*, 14,1080-1100, 1986.

[19] R. Rosales and S. Sclaroff. 3d trajectory recovery for tracking multiple objects and trajectory guided recognition of actions. In *CVPR*, 1999.

[20] R. Rosales and S. Sclaroff. Inferring body pose without tracking body parts. In *CVPR*, 2000.

[21] H. Siddenbladh, M. Black, and D. Fleet. Stochastic tracking of 3d human figures using 2d image motion. In *ECCV*, 2000.

[22] Y. Song, X. Feng, and P. Perona. Towards detection of human motion. In *CVPR*, 2000.

[23] C. Wren, A. Azarbayejani, T. Darrell, and A. Pentland. Pfinder: Real time tracking of the human body. *PAMI*, 19(7):780-785, 1997.

Figure 2. Example reconstruction of several testing sequences, each set (4 rows each) consists of input images, reconstruction using case (1), reconstruction using case (2a), and ground-truth, shown every 25th frame. View angles are 0 and $12\pi/32$ radians respectively for each set. Note that these sequences have challenging configurations, body orientation is also recovered correctly.

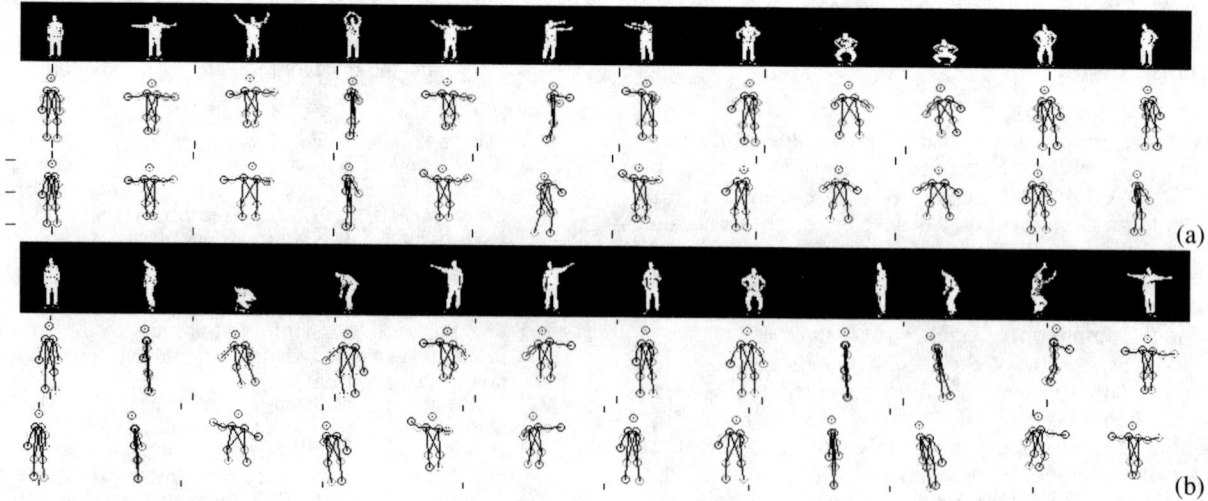

Figure 4. Reconstruction obtained from segmenting a human subject (every 30th frame). Two sequences are shown, each consists of input sequence, case (1) and case (2a) reconstructions

TALKING HEADS: Introducing the tool of 3D motion fields in the study of action

Jan Neumann and Yiannis Aloimonos
Center for Automation Research
University of Maryland
College Park, MD 20742-3275, USA
{jneumann,yiannis}cfar.umd.edu

Abstract

We demonstrate a method to compute three-dimensional (3D) motion fields on a face to serve as an intermediate representation for the study of actions. Twelve synchronized and calibrated cameras are positioned all around a talking person and observe its head in motion. We represent the head as a deformable mesh, which is fitted in a global optimization step to silhouette-contour and multi-camera stereo data derived from all images. The non-rigid displacement of the mesh from frame to frame, the 3D motion field, is determined from the normal flow information in all the images. We integrate these cues over time, thus producing a spatio-temporal representation of the talking head. Our ability to estimate 3D motion fields points to a new framework for the study of action. Using multi-camera configurations we can estimate a sequence of evolving 3D motion fields representing specific actions. Then, by performing a geometric and statistical analysis on these structures, we can achieve dimensionality reduction and thus come up with powerful representations of generic human action.

1 Introduction

What does it mean to understand an action? *One understands an action if one is able to imagine performing an action with images that are sufficient for serving as a guide in actual performance.* To be able to visualize or virtualize an action in our mental theater, we have to develop a spatio-temporal action description of the object in space that is performing the action. What are the key points in figuring out the nature of action representations?

1. Action representations are view independent. We are able to recognize and visualize actions regardless of viewpoint.

2. Action representations capture dynamic information which is manifested in a long image sequence. Put simply, it is not possible to understand an action on the basis of a small sequence of frames (viewpoints).

3. Action representations are made up of a combination of shape and movement.

To gain insights on action representations we consider them in a hierarchy with two more kinds of representations and the properties of the possible mappings among them. First there is the image data, that is, videos of humans in action. Considering the cue of motion, then our image data amounts to a sequence of normal flow fields computed from the videos. The second kind of representations are intermediate descriptions encoding information about 3D space and 3D motion, estimated from the input (video). These representations consist of a whole range of descriptions of different sophistication encoding partially the space-time geometry, and they are view and scene dependent. Finally, we have the action representations themselves, which are view and scene independent. An intermediate representation for the specific action in view is then a sequence of evolving 3D motion fields (also known as *range* [16] or *scene flow* [18]) and it is the most sophisticated intermediate description that could be obtained. Acquiring this representation is no simple matter, but it can be achieved by employing a very large number of viewpoints (e.g., for a general overview about human motion modeling see [1] and [9]).

As an example for an interesting action, we will examine facial expressions. Several image sequences of a talking and moving head were simultaneously recorded by a large number of cameras. From these image sequences a three-dimensional mesh model of the head was constructed and the trajectories of the mesh vertices in space-time, the evolving motion fields, were determined.

Due to the number of possible applications for spatio-temporal descriptions, for example in the field of human-computer-interaction or in entertainment (e.g., "Motion Capturing"), lot of work has been done on how to build 3D models of faces and how to synthesize and recognize facial expressions. Most approaches made only use of a few viewpoints at a time, thus they were not utilizing all the available constraints and information. For example, [8] and [14] fitted a predefined animation model to image data from few views and [20] used a single image in an analysis-by-synthesis loop.

Other methods needed a complicated prior motion and face model (e.g.,[17] and [6] use a physics-based model with anatomically correct muscles) or tracking markers on the face (.e.g, see [10]) to extract the facial expressions. The difference of our approach is that we construct a full three-dimensional model without manual intervention and animate them without relying on any prior model. The estimation of the 3D motion flow on the head surface is computed directly from image derivatives alone. Stereo and motion estimation were combined into one framework such as in [21] and [13], but in contrast to our approach the scene is still parameterized in the image space of a base view instead of the more natural object space parameterization. By moving the representation from image to object space, the algorithm can handle arbitrary camera arrangements and can make use of robust regularization constraints on the object surface, because physical tissue deforms in a continuous and smooth manner. The use of multi-camera setups for the computation of full 3D flow has only recently becoming feasible due to sinking costs of image capture and computer equipment (for an example see [19]).

To be able to to build up scene-independent representations for facial expressions, it is essential to separate the 3D motion flow field into a component due to a change of pose and a component due to the facial expression. Former approaches used simplified models such as planar models plus parallax for the head motion and affine motion models the facial expressions (e.g., [2]and [3]). Using the changing silhouettes and the rigid surface regions of the object to determine the rigid motion, we can compensate for the change in pose. After subtracting the rigid motion flow component from the full flow, we are left with the non-rigid residual motion describing the facial expression that can be analyzed or used for reanimation of other models.

2 Preliminaries and Definitions

We have established in our laboratory a multi-camera network consisting of sixty-four cameras, Kodak ES-310, providing images at a rate of up to eighty-five frames per second; the video is collected directly on disk –the cameras are connected by a high-speed network consisting of sixteen dual processor Pentium 450s with 1 GB of RAM each [5]. The camera configuration is parameterized by the camera positions $\mathbf{T}_k$, the rotation matrices R_k that relate the camera coordinate system to the fiducial system, and the intrinsic camera parameters K_k (in this paper bold-face letters denote vectors, small letters scalars, and large letters matrices). The calibration is done using images of a large calibration object. In the following we assume that the images have already been corrected for radial distortion. The image formation process is described by the conventional pinhole camera model, where the point $\mathbf{P}$ in fiducial world coordinates is related to its projection $\mathbf{p}_k$ in camera k as follows ($\mathbf{\hat{z}} = [0\,0\,1]^{\mathsf{T}}$):

$$\mathbf{p}_k = K_k \frac{R_k(\mathbf{P} - \mathbf{T}_k)}{\mathbf{\hat{z}} \cdot R_k(\mathbf{P} - \mathbf{T}_k)} \qquad (1)$$

The head surface is approximated by a closed mesh with vertices $\mathbf{V}_i$ and triangular facets $\mathbf{F}_j$. The world coordinates of $\mathbf{V}_i(t) = [x_i(t), y_i(t), z_i(t)]$ are dependent on the time t. Since we formulate the structure and motion estimation in object space, the image information needs to be sampled in regular patterns on the mesh surface instead of in regular patterns on the images. Therefore, a set of regularly spaced sampling points is associated with each triangle. The number of sampling points is dependent on the visible area of the triangle in the different cameras.

It is assumed that the head is the only moving object in all the image sequences, although this assumption is not essential and can be eliminated by applying the algorithm in turn to each independently-moving object. The following sections describe the algorithm that computes the spatio-temporal representation of the moving and talking head (from now on called the "object"):

- Section 3: Motion-based segmentation of the input images to locate the moving object, compute its silhouettes, and initialize the deformable 3D mesh.

- Section 4: Multi-camera stereo refinement of the deformable mesh where the search space is constrained by the silhouettes.

- Section 5: Computation of the 3D motion field on the mesh surface from image derivatives based on the normal flow constraint.

3 Image Segmentation

We incrementally construct an image of the background by modeling the temporal dynamics of the changing foreground pixels and the static background pixels. The magnitude of the temporal image derivatives and image statistics such as mean and variance are computed for each pixel on the surrounding 10 frames in the sequence and then used to segment the image into fore- and background. We integrate information over time to make the segmentation more robust by applying order-statistic filters over small spatio-temporal volumes. After the initial segmentation , we intersect the cone-shaped spaces formed by reprojecting the convex hulls of the head silhouettes into space. The intersection is a convex approximation of the head and it defines the initial 3D surface mesh. The mesh is now back-projected into each image and the segmentation is refined by fitting the mesh to all silhouette contours simultaneously.

4 Multi-Camera Stereo Estimation

Using information from silhouettes alone, it is not possible to compute more than the visual hull [12] of the object in view. Therefore, to refine our 3D surface estimate of the object, we adapt the vertices of the mesh to optimize the correlation between corresponding image regions in the different camera views. The search range for the vertex positions is constrained by the displacement boundaries computed in the silhouette estimation step in Section 3. To determine the visibility of each triangle, a z-buffer algorithm computes the index of the closest triangle patch for each pixel location. Next, a regular sampling point pattern is assigned to each mesh triangle as described before in Section 2, so that the sampling density of the closest image is about one projected sampling point per pixel.

We optimize orientation and position of each triangle by displacing each triangle vertex along the surface normal direction of the mesh and maximizing a similarity criterion among the triangle projections. The criterion to be optimized is the normalized cross-correlation between the projections of each triangle into all the cameras in which the triangle is visible (we denote this set of cameras as the set of "visible cameras"). For all combinations of normal displacements of the three vertices we compute the 3D coordinates of the sampling points on the triangle surface and project the sampling points into all the visible cameras. The image brightness of a projected sampling point is determined by bilinear interpolation. The cross-correlation is now computed between the corresponding image brightness

samples for all pairs of cameras that mutually see the triangle. We combine the correlation scores from all the camera pairs by taking a weighted average with the weights depending on the angle between camera plane and triangle plane.

The pairwise scores between all the cameras are also used to correct the visibility information. If a bimodal distribution of high and low correlation scores can be detected, then it is possible to estimate which cameras are visible and which are not, and the occluded cameras can be excluded from the score. For each vertex we collect the normal displacements corresponding to the highest correlation score for each of the surrounding triangles and determine the final normal displacement subject to global smoothness and rigidity constraints that are added to regularize the solution.

5 Motion Estimation

Following the description of the photometric properties of a surface in space in [11] and [18], the head surface is assumed to have Lambertian reflectance properties, thus the brightness intensity of a pixel $\mathbf{p}_k$ in camera k is given by

$$ I(\mathbf{p}_k; t) = -c_k \cdot \rho(\mathbf{P}) \cdot [\mathbf{n}(\mathbf{P}; t) \cdot \mathbf{s}(\mathbf{P}; t)] \qquad (2) $$

with an albedo $\rho(\mathbf{P})$ that is constant over time ($d\rho/dt = 0$) and where c_k is the constant that describes the brightness gain for each camera, $\mathbf{n}$ is the normal to the surface at $\mathbf{P}$, and $\mathbf{s}$ the direction of incoming light. Taking the derivative with respect to time on both sides, we get the following expression for the change of the image brightness $I(\mathbf{p}_k)$ at pixel location $\mathbf{p}_k$ in camera k:

$$ \frac{dI(\mathbf{p}_k)}{dt} = \nabla I(\mathbf{p}_k)^{\mathsf{T}} \cdot \frac{d\mathbf{p}_k}{dt} + \frac{\partial I(\mathbf{p}_k)}{\partial t} $$
$$ = -c_k \cdot \rho(\mathbf{P}) \cdot \frac{d}{dt}[\mathbf{n}(\mathbf{P}; t) \cdot \mathbf{s}(\mathbf{P}; t)] \qquad (3) $$

Since our sequences were recorded with a frame rate of 60 Hz and under fixed illumination, we can assume that $\frac{d}{dt}[\mathbf{n} \cdot \mathbf{s}] = 0$ and we end up with the well-known *normal flow constraint* equation.

$$ -\frac{\partial I(\mathbf{p}_k)}{\partial t} = \nabla I(\mathbf{p}_k)^{\mathsf{T}} \cdot \frac{d\mathbf{p}_k}{dt} \qquad (4) $$

This equation gives us one constraint per measurement, we can only determine the component of the optic flow that is normal to the image gradient, the normal flow. The estimation of the tangential flow along the iso-brightness contour is ill-posed. Regularizing the problems by imposing image-based smoothness conditions on the solution to equation (4) introduces artifacts at depth discontinuities and biases due to inhomogeneous gradient distributions [7].

Each normal flow vector in an image constrains the projection of the 3D motion flow to lie along a line parallel to the iso-brightness contour in the image, the normal flow constraint line. Thus the 3D motion flow vector has to lie on the plane defined by the normal flow constraint line and the optical center of the camera. The component of the 3D motion along the iso-brightness contour on the object surface is not recoverable. This is the aperture problem revisited in 3D. Nevertheless, if we assume that neighboring patches on the surface will move in an elastic manner, we can impose smoothness constraints on the motion of neighboring points. This smoothness assumption is physically justified as long as our mesh model has the same topology as the object in view, because nearly all real materials deform elastically when strain is applied.

The mesh representation of the head defines a correspondence map between the cameras and the full 3D motion flow at each mesh vertex is determined by combining the information from all the sampling points of the triangles neighboring the mesh vertex. To relate image derivatives and 3D motion flow using the normal flow constraint, we have to determine the Jacobian of the image formation equation (1) (R_3 is third row of matrix R and $K, R, \mathbf{T}$ refer to the calibration parameters of camera k):

$$
\begin{aligned}
\frac{d\mathbf{p}_k}{dt} &= \frac{\partial \mathbf{p}_k}{\partial \mathbf{P}_k}\frac{\partial \mathbf{P}_k}{\partial t} \\
&= \frac{\partial}{\partial t} K \frac{R(\mathbf{P}-\mathbf{T})}{R_3^\mathsf{T}(\mathbf{P}-\mathbf{T})} \\
&= \left(\frac{KR - \mathbf{p}_k R_3^\mathsf{T}}{R_3^\mathsf{T}(\mathbf{P}-\mathbf{T})} \right) \frac{\partial \mathbf{P}}{\partial t}
\end{aligned}
\tag{5}
$$

The derivative images are sampled at all locations where the sampling points associated with each triangle are visible. Let a given triangle of the mesh be defined by the vertices $\mathbf{V}_1, \mathbf{V}_2, \mathbf{V}_3$, then for each sampling point $P = \sum_{j=1,2,3} \lambda_j \mathbf{V}_j$ of this triangle we get the following constraint equation for each measurement:

$$
-\frac{\partial I(\mathbf{p}_k)}{\partial t} = \sum_{j=1,2,3} \lambda_j \left(\nabla I(\mathbf{p}_k) \cdot \frac{KR - \mathbf{p}_k R_3^\mathsf{T}}{R_3^\mathsf{T}(\mathbf{P}-\mathbf{T})} \right) \frac{\partial \mathbf{P}_j}{\partial t}
\tag{6}
$$

There is one equation per sampling point per visible image. To integrate these constraints, we stack these equations to form the $m \times n$ matrix L where m is the number of sampling points over all the triangles and their projections into all the visible cameras and n the number of vertices of the mesh times the three spatial dimensions. The matrices for the models presented are on the order of $100\,000 \times 3000$. To regularize the solution we add smoothness constraints as extra rows to L.

Since it is computationally infeasible to solve this large system directly, we form the normal equations of the over-constrained system and solve them with a preconditioned conjugate gradient method with either the motion field of the previous frame or the solution to a rigid motion approximation as starting vectors. The second choice worked very well to initialize the optimization, because most parts of a human head move rigidly. The the magnitude of the residual non-rigid flow is used to segment the mesh into rigidly and nonrigidly moving areas. This enables us to separate the motion field into two parts, one due to the change of pose and one due to the expression on the face.

6 Results

For our experiments we used twelve cameras placed in a dome-like arrangement around the head of a person that was expressing surprise (Figures 1a, 1b, 1c). The face of the person was painted with lines to add gradients to regions with little texture. The mesh resolution was approximately 900 vertices and 1200 triangles evenly spread over the surface.

After the initial structure estimation stage of our algorithm, we are able to synthesize texture-mapped views of the head from arbitrary viewing directions (Figures 1d-1f). The textures, coming always from the least oblique camera with respect to a given triangle, were not blended together to demonstrate the good agreement between adjacent texture region boundaries (Figures 1d, 1e, 1f). This shows that the spatial structure of the head was recovered very well.

The 3D motion flow field for the current frame is computed and used to propagate the mesh to the next frame. The propagated mesh is refined by new stereo and silhouette data, before the next 3D motion flow field is computed, and the process is repeated. The 3D motion field shown in (Figures 1g-1l) was computed by integrating the 3D flows of frames 4 to 16.

The rigid motion flow was computed by parameterizing the 3D motion flow vectors by the instantaneous rigid motion $\partial \mathbf{P}/\partial t = \mathbf{v} + \boldsymbol{\omega} \times \mathbf{P}$, where $\mathbf{v}$ is the instantaneous translation, and $\boldsymbol{\omega}$ is the axis of rotation [11]. This parameterized flow field was then fitted to the normal flow information in the images. By subtracting the rigid motion flow from the full flow, we extract the non-rigid flow. We can see that the rigid motion part (the turning of the head to the upper left) is recovered well, as the magnitude of the residual non-rigid flow on the forehead, nose and ears in Figure 1i is insignificant.

The non-rigid motion is also computed accurately, as we can easily see in the close up of the mouth region (Figure 1l) how the lips move forward, the mouth opens slightly, and the skin of the jaw stretches towards the chin.

7   Conclusion and Future Work

We presented an algorithm that computes an accurate spatio-temporal description of a non-rigidly moving human head. The description consists of the spatio-temporal trajectories of the mesh vertices, the evolving motion fields.

To see how these motion fields can be used, let us now consider the mapping from the 3D motion fields to the scene independent action representations. This mapping should be such that it extracts from a specific action quantities of a generic character common to all actions of the same type. These quantities most probably take the form of spatio-temporal patterns in four dimensions.

One way of obtaining such patterns is to perform statistics on a large enough sample (e.g., [15]). Considering, a particular action (e.g., talking or dancing), we can obtain data in the multi-camera laboratory described before for a large number of individuals. In each case we can obtain a 3D motion field and thus are able to build up a large data base of 3D motion fields. To this database a number of statistical techniques, such as principal component analysis, can be applied to reduce the dimensionality of this space and describe it with a small number of parameters. Another way of obtaining these patterns would be to study invariances related to symmetry, and geometric quantities in space-time (e.g., angles, velocities, accelerations, periodicity, etc. [4]).

For the future work, we will apply the above mentioned statistical and geometrical methods to the evolving 3D motion fields and try to extract the action representations. To improve the presented algorithm we plan to incorporate explicit visibility updating into the stereo part of the algorithm and include further information such as range flow constraints (see [16]) between the consecutive stereo reconstructions to estimate the 3D motion more accurately.

References

[1] J. Aggarwal and Q. Cai. Human motion analysis: A review. *CVIU*, 73(3):428–440, March 1999.

[2] B. Bascle and A. Blake. Separability of pose and expression in facial tracing and animation. In *ICCV98*, pages 323–328, 1998.

[3] M. Black and Y. Yacoob. Recognizing facial expressions in image sequences using local parameterized models of image motion. *IJCV*, 25(1):23–48, October 1997.

[4] O. Bottema and B. Roth. *Theoretical Kinematics*. North-Holland, 1979.

[5] L. Davis, E. Borovikov, R. Cutler, D. Harwood, and T. Horprasert. Multi-perspective analysis of human action. In *Proc. of Third International Workshop on Cooperative Distributed Vision*, Kyoto, Japan, 1999.

[6] I. Essa and A. Pentland. Coding, analysis, interpretation, and recognition of facial expressions. *PAMI*, 19(7):757–763, July 1997.

[7] C. Fermüller, R. Pless, and Y. Aloimonos. The Ouchi illusion as an artifact of biased flow estimation. *Vision Research*, 40:77–96, 2000.

[8] P. Fua and C. Miccio. Animated heads from ordinary images: A least-squares approach. *CVIU*, 75(3):247–259, September 1999.

[9] D. Gavrila. The visual analysis of human movement: A survey. *CVIU*, 73(1):82–98, January 1999.

[10] B. Guenter, C. Grimm, D. Wood, H. Malvar, and F. Pighin. Making faces. In *Proc. of SIGGRAPH*, pages 55–66, 1998.

[11] B. K. P. Horn. *Robot Vision*. McGraw Hill, New York, 1986.

[12] A. Laurentini. The visual hull concept for silhouette-based image understanding. *PAMI*, 16(2):150–162, February 1994.

[13] S. Malassiotis and M. Strintzis. Model-based joint motion and structure estimation from stereo images. *CVIU*, 65(1):79–94, January 1997.

[14] F. Pighin, J. Hecker, D. Lischinski, R. Szeliski, and D. Salesin. Synthesizing realistic facial expressions from photographs, 1998.

[15] D. Reynard, A. Wildenberg, A. Blake, and J. Marchant. Learning dynamics of complex motions from image sequences. In *ECCV96*, pages I:357–368, 1996.

[16] H. Spies, B. Jaehne, and J. Barron. Dense range flow from depth and intensity data. In *ICPR00*, pages xx–yy, 2000.

[17] D. Terzopoulos and K. Waters. Analysis and synthesis of facial image sequences using physical and anatomical models. *PAMI*, 15(6):569–579, June 1993.

[18] S. Vedula, S. Baker, P. Rander, R. Collins, and T. Kanade. Three-dimensional scene flow. In *ICCV99*, pages 722–729, 1999.

[19] S. Vedula, S. Baker, S. Seitz, and T. Kanade. Shape and motion carving in 6d. In *CVPR00*, pages II:592–598, 2000.

[20] T. Vetter and V. Blanz. Estimating coloured 3-d face models from single images: An example-based approach. In *ECCV98*, pages 499–513, 1998.

[21] Y. Zhang and C. Kambhamettu. Integrated 3d scene flow and structure recovery from multiview image sequences. In *CVPR00*, pages II:674–681, 2000.

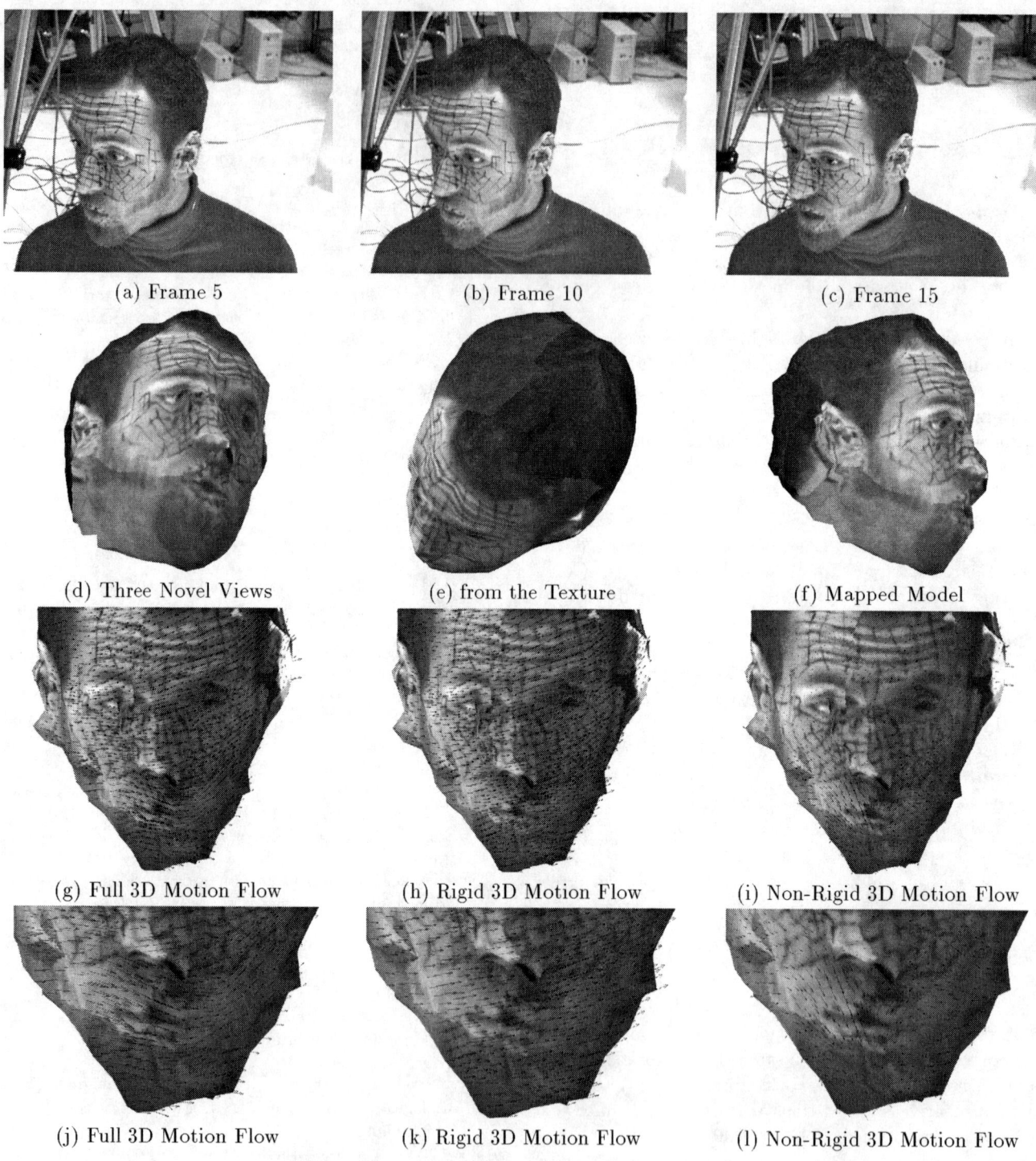

(a) Frame 5   (b) Frame 10   (c) Frame 15

(d) Three Novel Views   (e) from the Texture   (f) Mapped Model

(g) Full 3D Motion Flow   (h) Rigid 3D Motion Flow   (i) Non-Rigid 3D Motion Flow

(j) Full 3D Motion Flow   (k) Rigid 3D Motion Flow   (l) Non-Rigid 3D Motion Flow

Figure 1. Results of 3D Structure and Motion Flow Estimation

Motion Estimation & Recognition I

A Framework for Motion Recognition with Applications to American Sign Language and Gait Recognition

Christian Vogler, Harold Sun, and Dimitris Metaxas
Vision, Analysis, and Simulation Technologies Laboratory
Department of Computer and Information Science, University of Pennsylvania
200 S. 33rd Street
Philadelphia, PA 19104–6389
{cvogler,hsun}@gradient.cis.upenn.edu, dnm@central.cis.upenn.edu

Abstract

Human motion recognition has many important applications, such as improved human-computer interaction and surveillance. A big problem that plagues this research area is that human movements can be very complex. Manging this complexity is difficult. We turn to American Sign Language (ASL) recognition to identify general methods that reduce the complexity of human motion recognition.

In this paper we present a framework for continuous 3D ASL recognition based on linguistic principles, especially the phonology of ASL. This framework is based on parallel Hidden Markov Models (HMMs), which are able to capture both the sequential and the simultaneous aspects of the language. Each HMM is based on a single phoneme of ASL. Because the phonemes are limited in number, as opposed to the virtually unlimited number of signs that can be composed from them, we expect this framework to scale well to larger applications.

We then demonstrate the general applicability of this framework to other human motion recognition tasks by extending it to gait recognition.

1. Introduction

Human motion recognition is a field with a wide variety of applications. Of particular interest are gesture recognition for new modes of human-computer interaction, and gait recognition for video surveillance systems and intrusion detection. These applications share a common problem: Human movements can be very complex, with many actions taking place both sequentially and simultaneously. As an example of sequential complexity, consider a gesture that consists of a complex series of hand movements. As an example of simultaneous complexity, consider a human handing an object over to another human, while walking at the same time. Likewise, when a human performs a complex gesture, he could use both hands to perform two different actions at the same time.

Because there are so many different combinations of sequential and simultaneous human movement actions, it is impossible to model them all explicitly. We elaborate on this problem in Sec. 3.1 and Sec. 3.2. For this reason, a comprehensive framework for human motion recognition must provide a way to reduce the complexity of the problem. An obvious approach is to break down the actions into smaller primitives that are powerful enough to be combined into any conceivable action. Unfortunately, we have little data on what these primitives are for most human motion recognition applications, because they are relatively unconstrained.

American Sign Language (ASL) recognition yields valuable insights into the problem of managing complexity. Unlike most other motion recognition applications, ASL recognition is highly structured and constrained, thanks to the status of ASL as a language. Furthermore, the linguistics of ASL have been extensively researched (e.g., [13]), which helps us identify the primitives ("phonemes") of ASL. For this reason, it is beneficial to research ASL recognition first before applying the results to other research areas.

In this paper we describe a novel and extensive framework for continuous ASL recognition based on an extension to hidden Markov models (HMMs). The main contributions of this work are (1) modeling each sign in terms of its constituent phonemes, thus handling sequential complexity; (2) reducing simultaneous complexity by modeling signs in terms of independent channels and recognizing them with parallel HMMs, which are essentially regular HMMs applied to several channels simultaneously; and (3) recognizing signed sentences from full-fledged 3D data, which we collect either with a magnetic tracking system, or with 3D computer vision methods [7].

To demonstrate that the ASL recognition framework can be generalized, we discuss its application to gait recognition. Although gait and ASL are two very different areas,

33

we show that many concepts are similar. These similarities allow us to carry over the framework with virtually no modifications.

The rest of the paper is organized as follows: We discuss related work, then provide an overview on the phonological structure of ASL and its sequential and simultaneous aspects. We then describe how to model these aspects with parallel HMMs and provide experiments to verify our approach. We then generalize the framework to gait recognition. In the concluding remarks we discuss briefly what the framework has accomplished and provide an outlook for future work.

2. Related Work

Much previous work has focused on isolated sign language recognition with clear pauses after each sign, although the research focus is slowly shifting to continuous recognition. These pauses make it a much easier problem than continuous recognition without pauses between the individual signs, because explicit segmentation of a continuous input stream into the individual signs is very difficult. For this reason, and because of coarticulation effects, work on isolated recognition often does not generalize easily to continuous recognition.

Some isolated recognition work used neural networks [3, 16]. Other work focused on computationally inexpensive methods [6].

Most work on continuous sign language recognition is based on HMMs, which offer the advantage of being able to segment a data stream into its constituent signs implicitly. It thus bypasses the difficult problem of segmentation entirely.

T. Starner and A. Pentland used a view-based approach with a single camera to extract two-dimensional features as input to HMMs with a 40-word vocabulary and a strongly constrained sentence structure [12]. They assumed that the smallest unit in sign language is the whole sign. This assumption leads to scalability problems, as vocabularies become larger.

H. Hienz and colleagues used HMMs to recognize a corpus of German Sign Language [5] with 2D-based methods. They also experimented with stochastic bigram language models to improve recognition performance. The results of using stochastic grammars largely agreed with our results in [14].

Y. Nam and K. Y. Wohn [10] used three-dimensional data as input to HMMs for continuous recognition of gestures. They introduced the concept of movement primes, which make up sequences of more complex movements. The movement prime approach bears some superficial similarities to the phoneme-based approach in this paper.

R. H. Liang and M. Ouhyoung used HMMs for continuous recognition of Taiwanese Sign Language with a vocabulary between 71 and 250 signs. [8] Unlike other work in this area, they did not use the HMMs to segment the input stream implicitly. Instead, they segmented the data stream

explicitly based on discontinuities in the movements. They integrated the handshape, position, orientation, and movement aspects at a higher level than the HMMs.

We used HMMs and 3D computer vision methods to model phonological aspects of ASL with an unconstrained sentence structure [14]. In [15] we extended the conventional HMM framework to capture the parallel aspects of ASL, which ordinarily would make the recognition task too complex.

3. Overview of the Framework

We now discuss the two main aspects of our framework, which are the linguistic modeling of ASL, and the modeling of the sequential and simultaneous aspects of ASL with a novel HMM-based approach. Although taking advantage of research into linguistics to model the signs is specific to signed languages, the principal idea of breaking down larger units into their constituent parts applies to other recognition tasks. Likewise, the HMM framework can be applied to other recognition tasks without alterations. In Sec. 5 we show by example of gait recognition how to extend our framework to other applications.

3.1. ASL Linguistics

ASL is the primary mode of communication for many deaf people in the USA. It is a highly inflected language; that is, many signs can be modified to indicate subject, object, and numeric agreement. They can also be modified to indicate manner (fast, slow, etc.), repetition, and duration [13]. Like all other languages, ASL has structure, which sets it clearly apart from most other human motion recognition problems. It allows us to test ideas in a constrained framework first, before attempting to generalize the results.

In particular, managing the complexity of large data sets is an area where ASL recognition work can yield valuable insights. Managing complexity is already difficult in the constrained field of ASL recognition, because signs can appear in many different forms, both sequentially and simultaneously. Other human motion recognition applications are often much less constrained than ASL, so this problem will only be exacerbated. It is, therefore, important to develop methods that make the complexity of ASL, and, by extension, other human motion recognition problems manageable.

The key idea behind managing complexity is that actions can be broken down into smaller subunits, and that any action can be described in terms of these subunits. In the case of ASL these subunits are called **phonemes**[1]. Formally, a phoneme is defined to be the smallest contrastive unit in a language. In English, examples of phonemes are the sounds $/c/$, $/a/$, and $/t/$. In ASL, examples of phonemes are the

---

[1]Some people prefer to associate the term "phoneme" with spoken languages only, and use the term "chereme" for sign languages. We follow the terminology of spoken language linguistics, because the underlying concepts are the same.

34

Figure 1. The sign for "father." The white X indicates contact between the thumb and the forehead after each tap. The location of the hand at the forehead and the tapping movements are examples of phonemes.

movement of the hand toward the chin in the sign for "FATHER," and the starting location of the hand in front of the forehead at the beginning of this sign (Fig. 1).

Phonemes are *limited in number,* as opposed to the *virtually unlimited number* of words or signs that can be constructed from them. In English, there are approximately 40 distinct phonemes, whereas in ASL there are approximately 150–200 distinct phonemes[2]. For this reason, taking advantage of phonology can make an otherwise intractable modeling task feasible. It is practical to provide enough training data for a small set of phonemes, from which every sign can be constructed. Doing the same for signs that are not modeled in terms of phonemes would become impossible with vocabularies larger than a few hundred signs.

Unlike in spoken language linguistics, sign language linguists have not yet agreed on a common phonological model for ASL. Surveying all of the different phonological models is beyond the scope of this paper. We now briefly describe the one that we use in our recognition framework.

The Movement-Hold Model The Movement-Hold model [9] assumes that signs can be broken down into two major types of segments, which are called **movements** and **holds**. Movements are those segments, during which some aspect of the signer's configuration changes, such as a change in handshape, or a hand movement from one location to another. Holds, in contrast, are those segments, during which the hands remain translationally stationary.

Signs are made up of sequences of movements and holds. A very common sequence is *MMMH* (three movements followed by a hold), such as in the sign for "FATHER" (Fig. 1). This sign starts out with a movement toward the forehead, then away from the forehead, toward the forehead again, followed by a hold touching the forehead. Attached to each segment is a bundle of articulatory features, which primar-

---

[2]This number applies to the the Movement-Hold phonological model [9] described in Section 3.1. The numbers for other models vary slightly.

ily describe the handshape, orientation, and location of each segment. Fig. 2 shows a schematic example.

For a detailed description of all the existing phonemes in the Movement-Hold model, see [9]. For a detailed description of the phonemes that we have used so far in our framework, see [15].

Simultaneous Aspects of ASL The Movement-Hold model is ideally suited for ASL recognition, because it emphasizes sequential aspects over simultaneous aspects. This emphasis fits HMMs very well, because they are sequential in nature. Yet, despite the emphasis on sequentiality, a lot of phonemes also occur simultaneously. For example, often the handshape changes simultaneously with the hand movement in a sign. Likewise, many signs are two-handed, and both hands move simultaneously. A purely sequential framework cannot capture this kind of simultaneity.

A look at the Movement-Hold model immediately suggests an approach to incorporating simultaneity into the framework by modeling all possible combinations of segments and feature bundles. This approach fails because of the sheer number of possible combinations of phonemes. If we consider both hands, and assume 30 basic handshapes, 8 hand orientations, 8 wrist orientations, and 20 major body locations [9], the total number of phoneme combinations is $(30 \times 8 \times 8 \times 20)^2 \approx 1.5 \times 10^9$. Even if we employ some constraints on the weak hand for two-handed signs, the number is still approximately 2.9×10^8 [15]. It would be impossible to get enough training data for 10^9 models.

This problem is not unique to sign language recognition. Many other motion recognition applications, such as gestures and full human body movement, are even worse off, because they are less constrained than ASL. For this reason, a different approach toward handling simultaneous processes is necessary.

For this reason, we make a major modification to the Movement-Hold model. Instead of attaching the bundles of articulatory features to the movement and hold segments,

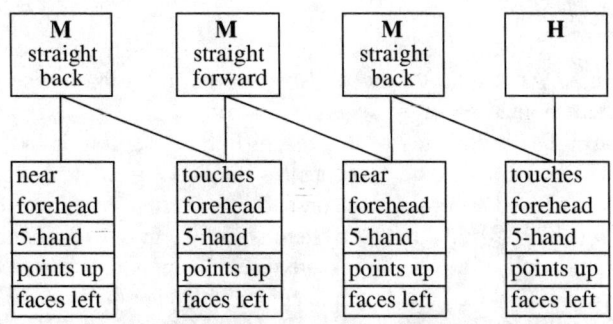

Figure 2. Schematic description of the sign for "FATHER" in the Movement-Hold model. It consists of three movements, followed by a hold (compare Fig. 1).

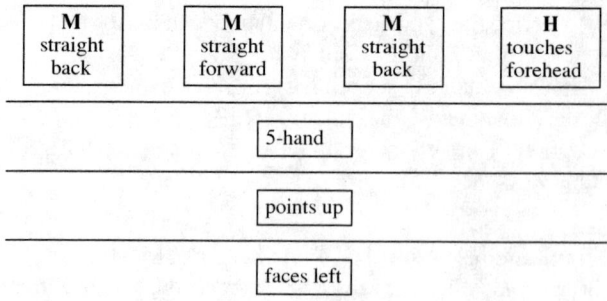

| M straight back | M straight forward | M straight back | H touches forehead |

5-hand

points up

faces left

Figure 3. The sign for "father," where the different features are modeled in separate channels. Compare with Fig. 2.

we break them up into **channels** that are independent from one another. One channel consists of movements and hold segments that describe the type of movement and the body locations of the right ("strong") hand. Other channels could consist of the segments in the left ("weak") hand, the handshape, the hand orientation, and the wrist orientation. Figure 3 shows how the sign for "FATHER" is represented with this modification.

By modeling the simultaneous aspects of ASL as independent channels, we gain the ability to model each channel separately, yet combine each channel on the fly during recognition of a signed sentence. The success of this approach depends primarily on how independent the channels are from one another in reality. In the case of ASL, there is some linguistic evidence that the strong and weak hands move independently from each other [9]. Our experiments in Sec. 4 suggest that the independence assumption is at least partially valid.

3.2. Recognition with Hidden Markov Models

One of the main challenges in ASL recognition is to capture the variations in the signing of even a single human. Hidden Markov models (HMMs) are a type of statistical model embedded in a Bayesian framework and thus well suited for capturing these variations. In addition, their state-based nature enables them to describe how a signal changes over time.

An HMM λ consists of a set of N states $S_1, S_2, \ldots, S_N$. At regularly spaced discrete time intervals, the system transitions from state S_i to state S_j with probability a_{ij}. The probability of the system initially starting in state S_i is π_i. Each state S_i generates output $O \in \Omega$, which is distributed according to a probability distribution function $b_i(O) = P\{\text{Output is } O | \text{System is in } S_i\}$. In most recognition applications $b_i(O)$ is a mixture of Gaussian densities.

We use one HMM per phoneme, which are then chained together to form the signs. The individual signs in turn are chained together into a network. Then the recognition problem is reduced to finding the most likely state sequence through the network that could have generated the input signal with the signs to be recognized. From the state sequence

the sequence of signs, and hence the recognized sentence, can be recovered.

The Baum-Welch algorithm is used to train HMMs on a set of training data in polynomial time, and the Viterbi algorithm is used to find the most likely state sequence in polynomial time through a network of HMMs during the recognition phase. For details on these algorithms, see [11].

Modeling Simultaneous Aspects Regular HMMs, as we have described them so far, can model the sequential aspects of the Movement-Hold model well, but they are not suitable for modeling the simultaneous aspects. In the past, researchers have suggested using factorial hidden Markov models [4] and coupled hidden Markov models [2]. Although these two approaches are good at capturing simultaneous, coupled processes, they would still require *a priori* knowledge of all the possible phoneme combinations at training time. In other words, they do not address the underlying problem, which is the sheer number of possible phoneme combinations.

Instead, we introduce Parallel HMMs (PaHMMs) as a modification to the HMM framework that directly reflects the decomposition of the simultaneous aspects of ASL into independent channels, as described in Sec. 3.1. We model each channel separately with HMMs and train them separately. At recognition time, PaHMMs combine the probabilities from each channel by multiplying them. That is, PaHMMs are essentially regular HMMs that are used in parallel. We describe the details of this approach and the algorithms for PaHMMs in [15].

Because the channels are independent, the complexity problems with the number of possible phoneme combinations disappear. With PaHMMs we can train the phonemes in each channel separately and put together new, previously unseen combinations of phonemes on the fly. Thus, during the training phase, we need only enough data for a robust estimate of the HMMs' parameters for each phoneme, instead of all combinations of these.

4. Experiments

We ran several continuous recognition experiments with 3D data to test the feasibility of modeling the movements of the left and the right hands with PaHMMs. We used two channels, which modeled the movements and holds of the left and the right hands, respectively. Our database consisted of 400 training sentences and 99 test sentences over a vocabulary of 22 signs. The transcriptions of these signs are listed in [15].

We collect the sentences with an Ascension Technologies MotionStar™ 3D tracking system, and with our vision-based tracking system at 60 frames per second. The latter uses physics-based modeling to track the arms and the hands of the signer, as depicted in Figure 4. The physics-based models are estimated from the images from a subset of three orthogonal cameras. These are selected on a

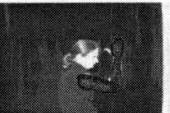

Figure 4. These images show the 3D tracking of the sign for "father."

Regular HMMs

Level	Accuracy	Details
sentence	80.81%	H = 80, S = 19, N = 99
sign	93.27%	H=294, D=3, S=15, I=3, N=312

Parallel HMMs

Level	Accuracy	Details
sentence	84.85%	H = 84, S = 15, N = 99
sign	94.23%	H=297, D=3, S=12, I=3, N=312

Table 1. Results of the recognition experiments. H denotes the number of correct sentences or signs, D the number of deletion errors, S the number of substitution errors, I the number of insertion errors, and N the total number of sentences or signs in the test set.

per-frame basis depending on the occluding contour of the signer's limbs [7].

We used an 8-dimensional feature vector for each hand. Six features consisted of 3D positions and velocities relative to the base of the signer's spine. For the remaining two features, we computed the largest two eigenvalues of the positions' covariance matrices over a window of 15 frames centered on the current frame. In normalized form, these two eigenvalues provide a useful characterization of the global properties of the signal.

In the experiments we compared the recognition accuracy of modeling only the movements and holds of the right hand with regular HMMs and modeling both hands with PaHMMs. The results are given in Table 1 and show that one the sentence level, the difference in recognition accuracy between regular and parallel HMMs is significant. Hence, PaHMMs can make the recognition system more robust.

5. Extensions to Gait Recognition

Most of the framework for ASL recognition readily carries over to gait recognition. To test this hypothesis, we set up an experiment within our framework to discriminate among walking on level terrain, walking upward a slope, and walking downward a slope.

The basic unit in gait recognition is the half-step; that is, the time a leg takes to complete one of the stance or swing phases. A step consists of two half-steps. The first half-step

Pelvis segment and y–axis   Pelvis elevation angle

Figure 5. Sagittal elevation angles. We calculate them from the 2D positions of the markers at the sites indicated on the pictures.

models the leg during the stance phase, and the second one models the leg during the swing phase. The type of gait can change any time a half-step has been completed. Thus, concepts from ASL recognition have direct equivalents in gait recognition: A whole signs correpsonds to a step, and a phoneme corresponds to a half-step.

Before describing the experiment, we briefly cover how to represent gait data, which is very different from ASL data.

5.1. Data Representation

Elevation angles measure the orientation of a limb segment with respect to a vertical line in the world. We define the limb segment $\vec{v}$ between two points $\vec{a}$ and $\vec{b}$ on the body: $\vec{v} = \vec{a} - \vec{b}$. Typically $\vec{a}$ and $\vec{b}$ are points at opposite ends of a limb. The **sagittal elevation angles** are obtained by first projecting $\vec{v}$ onto the sagittal plane to form $v^{\vec{s}ag}$. The angle between $v^{\vec{s}ag}$ and the negative y axis is its sagittal elevation angle, ψ (Fig. 5).

We have followed the definition of elevation angles and placement of markers as used in [1], with the addition of a heel marker. Unlike joint angles and absolute coordinate values of the limbs, elevation angles are invariant with respect to different size humans. In addition, they appear to be invariant across different humans, as long as they perform the same kind of walking activity (e.g., walking on a level plain, walking on a slope) [1]. This property makes elevation angles a compelling choice for recognition features, especially for person-independent gait recognition.

5.2. Experiment

The task of the experiment was to discriminate among walking on level terrain, walking upward on slopes, and walking downward on slopes; as well as to identify the timing of the half-steps correctly. The slopes had different inclinations anywhere between 8 and 15 degrees. The shape of the terrain affects only the elevation angle of the foot, whereas the other angles appear to be unaffected. For this reason, we used the three elevation angles of the lower leg, the upper leg, and the pelvis as the feature vector.

We measured the elevation angles from a walking subject with the help of markers, as shown in Fig. 5. Future work could use our framework for tracking 3D body models [7], instead, to measure the elevation angles from any perspective. For the training set we used a set of ten measurements from a single person for each of level terrain, a 15 degree upward slope, and a 15 degree downward slope. Hence, we used a total of six HMMs — two for each type of step — chained together into a network. The sampling rate was 60 frames per second.

The test set contained the elevation angles of a person walking across uneven terrain. The recognizer was able to identify all half-steps in the test set correctly. The recognition of the timing of the steps worked well, as long as the type of step did not change. At transitions from one type of step to another, the recognizer often identified the end of the half step up to seven frames too early or too late. One possible explanation is that the elevation angles behave differently during a transition. In this case, modeling the transitions explicitly with HMMs, similar to modeling transitions between signs in sign language recognition [14], might improve the results.

6. Conclusions

We have developed a framework for human motion recognition. Although we initially applied it to ASL recognition, we have shown by example of gait recognition that it can be generalized to other recognition tasks. This makes our framework a promising contribution to the areas of human-computer interaction and video surveillance tasks.

Future work in ASL recognition should model other channels, such as handshape and orientation, and incorporate facial expressions, which constitute a large part of the grammar of ASL. It should also verify the framework with larger vocabularies. However, a prerequisite to experimenting with large vocabularies is a standardized corpus of ASL sentences. No such corpus exists at present.

Future work in gait recognition should model the transitions between different types of steps, incorporate more different types of steps (e.g., climbing a ladder or a stair), and model the differences between walking and running. It should also use 3D human body tracking, instead of measuring the sagittal elevation angles from the side with the help of markers.

Acknowledgments

This work was supported in part by an NSF Career Award NSF-9624604, ONR Young Investigator Proposal, NSF IRI-97-01803, AFOSR F49620-98-1-0434, and NSF EIA-98-09209.

References

[1] A. Borghese, L. Bianchi, and F. Lacquaniti. Kinematic determinants of human locomotion. *J. Physiology*, (494):863–879, 1996.

[2] M. Brand, N. Oliver, and A. Pentland. Coupled hidden Markov models for complex action recognition. In *Proceedings of the IEEE Conference on Computer Vision and Pattern Recognition*, 1997.

[3] R. Erenshteyn and P. Laskov. A multi-stage approach to fingerspelling and gesture recognition. Proceedings of the Workshop on the Integration of Gesture in Language and Speech, Wilmington, DE, USA, 1996.

[4] Z. Ghahramani and M. I. Jordan. Factorial Hidden Markov Models. *Machine Learning*, 29:245–275, 1997.

[5] H. Hienz, K.-F. Kraiss, and B. Bauer. Continuous sign language recognition using hidden Markov models. In Y. Tang, editor, *ICMI'99*, pages IV10–IV15, Hong Kong, 1999.

[6] M. W. Kadous. Machine recognition of Auslan signs using PowerGloves: Towards large-lexicon recognition of sign language. In *Proceedings of the Workshop on the Integration of Gesture in Language and Speech*, pages 165–174, Wilmington, DE, USA, 1996.

[7] I. Kakadiaris, D. Metaxas, and R. Bajcsy. Model based estimation of 3d human motion with occlusion based on active multi-viewpoint selection. In *Proceedings of the CVPR*, pages 81–87, 1996.

[8] R.-H. Liang and M. Ouhyoung. A real-time continuous gesture recognition system for sign language. In *Proceedings of the Third International Conference on Automatic Face and Gesture Recognition*, pages 558–565, Nara, Japan, 1998.

[9] S. K. Liddell and R. E. Johnson. American Sign Language: The phonological base. *Sign Language Studies*, 64:195–277, 1989.

[10] Y. Nam and K. Y. Wohn. Recognition and modeling of hand gestures using colored petri nets. To appear in IEEE transactions on Systems, Man and Cybernetics (A), 1999.

[11] L. R. Rabiner. A tutorial on Hidden Markov Models and selected applications in speech recognition. *Proceedings of the IEEE*, 77(2):257–286, 1989.

[12] T. Starner, J. Weaver, and A. Pentland. Real-time American Sign Language recognition using desk and wearable computer based video. *IEEE Transactions on Pattern Analysis and Machine Intelligence*, 20(12):1371–1375, 1998.

[13] C. Valli and C. Lucas. *Linguistics of American Sign Language: An Introduction*. Gallaudet University Press, Washington DC, 1995.

[14] C. Vogler and D. Metaxas. Adapting hidden Markov models for ASL recognition by using three-dimensional computer vision methods. In *Proceedings of the IEEE International Conference on Systems, Man and Cybernetics*, pages 156–161, Orlando, FL, 1997.

[15] C. Vogler and D. Metaxas. A framework for recognizing the simultaneous aspects of American Sign Language. To appear in *Computer Vision and Image Understanding*, 2001.

[16] M. B. Waldron and S. Kim. Isolated ASL sign recognition system for deaf persons. *IEEE Transactions on Rehabilitation Engineering*, 3(3):261–71, September 1995.

An Incremental Approach Towards Automatic Model Acquisition for Human Gesture Recognition

Michael Walter †   Alexandra Psarrou † and Shaogang Gong ‡
† Harrow School of Computer Science, University of Westminster,
Harrow HA1 3TP, U.K. `zeoec,psarroa@wmin.ac.uk`
‡ Dept. of Computer Science, Queen Mary and Westfield College,
London E1 4NS, U.K. `sgg@dcs.qmw.ac.uk`

Abstract

The recognition of natural gestures typically involves (a) the collection of training examples, (b) the generation of models and (c) the determination of a model that is most likely to have generated an observation sequence. The first step however, the collection of training examples typically involves manual segmentation and hand labelling of image sequences. This is a time consuming and labour intensive process and is only feasible for a limited set of gestures. To overcome this problem we suggest that gestures can be viewed as repetitive sequence of atomic movements, similar to phonemes in speech. We present an approach (1) to automatically segment an arbitrary observation sequence of a natural gesture, using only contextual information derived from the observation sequence itself and (2) to incrementally extract a set of atomic movements for the automatic model acquisition of natural gestures. Atomic components are modelled as Semi Continuous Hidden Markov Models and the search for repetitive sequences is done using a discrete version of CONDENSATION that is no longer based on factored sampling.

1. Introduction

Recognition of natural gestures typically involves three steps: (a) The collection of example gestures to be recognised, (b) the generation of models that capture the variations within the training examples and (c) the determination of a model that is most likely to have generated an observation sequence [3, 12, 13, 9]. Natural gestures, however, are expressive body motions with underlying spatial and in particular temporal structure which is probabilistic and rather ambiguous in nature. Problems are inevitably encountered in the segmentation and alignment of training examples due to measurement noise, nonlinear temporal scal-

ing based on variations in speed and most notably human variation in performing a gesture. Therefore segmentation in gesture recognition typically involves manual intervention and hand labelling of image sequences. It is clear that such an approach cannot meet the demands of many challenging recognition tasks, and is only feasible for a limited set of gestures.

In this paper we present an approach (1) to automatically segment an arbitrary observation sequence of a natural gesture, using only contextual information derived from the observation sequence itself and (2) to incrementally extract a set of atomic movements for the automatic model acquisition of natural gestures.

Our approach is based on the assumption that gestures can be viewed as a repetitive sequence of primitive movements, similar to phonemes in speech, starting and ending in a rest position and governed by a high level structure controlling the temporal sequence. The later is motivated by recent research in the field of Natural Gestures [10], that has identified five basic gesture types, iconic, metaphoric, cohesive, deistic and beat gestures. All gestures in common is their temporal signature. A hand gesture is typically embedded by the hands being in a rest state and can be divided into either bi-phasic or tri-phasic gestures. Beat and deictic gestures are examples for bi-phasic gestures. They have just two movement phases, away from the rest state into gesture space and back again, while iconic metaphoric and cohesive gestures have three, preparation, stroke and retraction. They are called tri-phasic and are executed first by transitioning from the rest state into gesture space (preparation), followed by a small movement (stroke), remaining in that configuration for a short duration and than returning back to the rest state (retraction).

This leads to our definition of a gesture as a recurrent sequence of atomic components, starting and ending in a rest position. The acquisition of models can now be described as follows: The first step is to build a 'phonetic' database of atomic components. The continuous observation sequence is automatically segmented into plausible atomic components. Those components are analysed for repetitions and the repeating sequences are incrementally stored in a database. In a second step the high level structure controlling the temporal sequence is determined. Rest positions are identified, the enclosed components analysed for repetitions and their temporal sequence used to build the gesture models.

The remaining part of this paper is organised as follows: Section 2 describes related work. Section 3 gives a system overview and describes the representation and segmentation of gestures and the encoding and recognition of atomic components. Section 4 describes our experiments. Section 5 concludes with a discussion and a perspective on future work.

2 Related Work

In psychology McNeil [10] identified five basic types of gestures, iconic, metaphoric, cohesive, deistic and beat gestures. Iconic gestures bear a close formal relationship to the semantic content of speech and refine the mental representation of a scene. ("And then he bent the tree"). Metaphoric gestures are similar to iconic gestures in that they are pictorial, but their contents present an abstract idea rather than a concrete object or event. This allows a metaphoric to create a concrete image of an abstract concept and present it to a listener ("She had a really nice shape"). Cohesive Gestures are to tie thematically related, but temporally separated parts together and thus can resume a story after it got interrupted. Deictic gestures are familiar pointing gestures. Their function is to point to objects and events in a concrete world or to point to metaphorical pictures or ideas in an abstract world and consequently giving them a physical location. Beat gestures seem to be some of the most insignificant gestures but reveal most about a speaker's conception of the narrative discourse. The value of beats lies in the fact that they index the words they accompany and emphasise them as being significant.

Approaches in Computer Vision for classification, learning and recognition of activities such as natural gestures and behaviour include:

Engel and Rubin [4] describe an approach for the qualitative classification of motion events. The events consist of smooth starts, smooth stops, pauses, impulse starts and impulse stops and are considered as motion events that partition a global motion into its psychological parts. Their method is based on a polar velocity representation and the derivation of first and second order derivatives.

Wilson et al [16] presented an approach for the qualitative classification of natural gestures into either bi-or tri-phasic gesture. They identify plausible rest-state configurations of a speaker telling a story and parse the sequences in between into either bi-or tri-phasic gestures using a priory knowledge of the temporal structure describing both gestures types.

Wilson and Bobick [17] describe an adaptive approach for unsupervised online learning of simple gestures for interactive control. Their algorithm requires a model of the temporal structure for the gesture to be learned, combined with contextual information derived from the application to bias the system in the early stages of runtime.

Vogler and Metaxas [14] present an approach to continuous, whole-sentence American Sign Language (ASL) recognition, based on a sequential phonological model of ASL. They break ASL into movements and holds, both are considered phonemes and subsequently train Hidden Markov Models to recognise the phonemes, instead of whole signs.

Galata et al [5] present an approach for the acquisition of statistical models of structured and semantically rich behaviour. Activities are modelled as sequence of atomic behaviour components, with variable length Markov models controlling the high level structure. Atomic behaviour components are seen as prototype sequence between two key prototypes, which in turn are identified as prototypes within the sequence, where changes drop below a preset threshold. The method can be used to generate and predict realistic human behaviour but can not generalise to previously unseen sequences.

Johnson [8] presented an approach for the automatic acquisition of statistical behaviour models from continuous observations of long image sequences and derived a method for the continuous assessment of behaviour typicality by exploiting the statistical nature of his behaviour models.

3 Incremental Learning of Natural Gestures: A combined approach

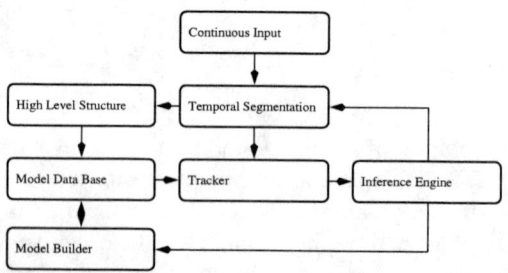

Figure 1. System architecture for the incremental learning approach.

Figure 1 gives an overview of the system architecture. The continuous input stream is passed to a segmentation unit (3.1) and automatically segmented into plausible atomic components. This involves the search for suitable break points and is done by analysing the observation sequence for potential rest positions and discontinuities, such as those that occur at transitions into and out of gesture space. The extracted segments are passed on to a tracker unit (3.6) and compared with the models stored in a model database (3.3). Previously unseen models are added to the database, thus incrementally extending it and repetitive sequences are used to refine the models. The sequence of atomic components is stored in a High Level Database and allows to analyse the observations for any high level structure controlling the temporal sequence.

3.1 Temporal Segmentation

Temporal segmentation serves the purpose to isolate primitive movements and to split the continuous sequence into plausible atomic segments. It is done in three stages. First the complete observation sequence is analysed for potential rest positions, which are identified as areas (3.2) where the motion drops below a threshold for a reasonable amount of time. A second step analyses the segments between two potential rest positions for short periods, where the velocity drops below a preset threshold, to recover pause positions that typically occur in tri-phasic gestures between stroke and retraction. A third step analyses the segments for discontinuities in the orientation of the tangent to the curve of the input trajectory using multi-scale analysis to find discontinuities such as transitions into and out of gesture space for bi-phasic gestures and discontinuities between preparation and stroke for tri-phasic gestures. The applied method is based on Asada and Brady's Curvature Primal Sketch [1], see Figure 2. The orientation of the tangent of a two dimensional hand trajectory $f(t)$ is convoluted with the first $N'_\sigma * f$ and second derivatives $N''_\sigma * f$ of a Gaussian $N_\sigma(t) = (1/(\sqrt{2\pi}\sigma))exp(-t^2/2\sigma^2)$ at different temporal scales $\sigma = \{\sigma_{min} \cdots \sigma_{max}\}$. The filter responses are analysed for characteristic maxima and zero crossings and only discontinuities consistent over a large scale are registered, thus taking care of noise at different scales.

3.2 Gesture Representation

Let us first define a gesture as an ordered sequence of states in a 2 dimensional spatio-temporal space. Each state k_i is defined by a set of parameters $< \mu_i, \Sigma_I >$ to capture the spatial information, where μ_i is the 2D centroid of a state and Σ_i the $2 X 2$ spatial covariance. Mean position and covariances are computed for the complete 2 dimensional observation sequence using dynamic k-means. The algorithm starts with a small number of states and continues to add new states until all the covariances drop below a preset

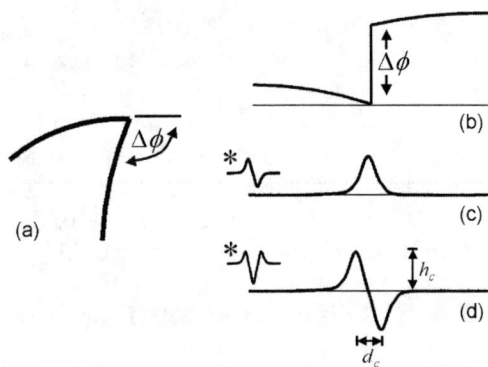

Figure 2. a) Example trajectory containing a curvature discontinuity $\Delta\Phi$. b) The trajectory in orientation space, relating the orientation of the tangent to the curve to the arc length along the curve. c) Filter response $N'_\sigma * f$ of the orientation of the trajectory $f(t)$ convoluted with the first derivative of a Gaussian $N_\sigma(t) = (1/(\sqrt{2\pi}\sigma))exp(-t^2/2\sigma^2)$. d) Filter response $N''_\sigma * f$ of the orientation of the trajectory $f(t)$ convoluted with the second derivative of a Gaussian $N_\sigma(t)$. As shown in d) corners give rise to a pair of peaks with a separation $d_c \approx 2\sigma$ and height $h_c \approx \frac{|\Phi|}{\sqrt{2e\pi}\sigma^2}$. Note, d_c is linearly dependent on the scale constant σ and monotonically decreases with σ, which provides a strong clue for the detection of corners.

threshold. Computing means and covariances has several purposes. It allows us to localise potential rest positions, to support the temporal segmentation and we can use it to generate a codebook for a Semi Continuous Hidden Markov Model (SCHMM)[6] to encode temporal information.

3.3 Encoding temporal information

Temporal information of gestures is encoded using Semi Continuous Hidden Markov Models. They were proposed to extend discrete HMMs [11] by replacing a set of discrete output probabilities with a set of continuous density function to combine the simplicity of discrete HMMs with the robustness of continuous distributions. They can be considered as a special form of continuous mixture HMMs with tied mixture coefficients to reduce the number of free parameters, training examples and computational complexity, while retaining the modelling powers of continuous HMMs.

The spatio-temporal structure of a gesture can be described as forward chained SCHMM , defined by a set of parameter $\lambda^m = (A^m, B^m, \pi^m)$ where: $A^m \in \{a^m_{ij}\}$ is a set of state transition probabilities for model m, describing the probability $p(q^m_{t+1} = j \,|\, q^m_t = i)$ of being in a state i at time t and state j at time $t+1$ where $\sum_{j=1}^{N} a^m_{ij} = 1$. $B \in \{b^m_j(o) = p(o_t|q^m_t = j)\}$ is an observation density describing the probability of an observation o_t in state $q^m_t = j$. It is composed as a Gaussian mixture $b^m_j(o) = \sum_{i=1}^{I} c^m_{ij} \mathcal{N}(o, \mu_i, \Sigma_i)$ with mixture coefficient c^m_{ij}, tied means μ_i and covariances Σ_i taken from the codebook for the ith mixture; $\pi^m \in \{\pi_1^m, \pi_2^m, \ldots \pi_N^m\}$, the initial probabilities of being in state i at time $t = 1$ where $\sum_{i=1}^{N} \pi_i^m = 1$.

The initial state probabilities are set to $\pi^m = \{1, 0, \ldots, 0\}$ according to a forward chained SCHMM, state transition probabilities A^m and mixture Coefficients C^m are computed using a modified version of the Baum-Welch algorithm. For more information see [11].

3.4 Semi CONditional DENSity PropagATION

Traditional approaches to recognise temporal structures based on Hidden Markov Models [18, 13, 12] apply the Forward-Backward Procedure or the Viterbi Algorithm. We take a different approach and apply a modified version of the Condensation algorithm [7]. Condensation has been successfully applied to gesture recognition [2, 15] and has the advantage to be able to propagate multiple hypothesis simultaneous in time. It can give instant information on the current model probability and moreover does not require the complete data segment to be provided.

Our approach is an extension to the method introduced in [15]. But unlike original CONDENSATION that works with continuous state distributions, works with semi continuous state distributions. The following two sections now

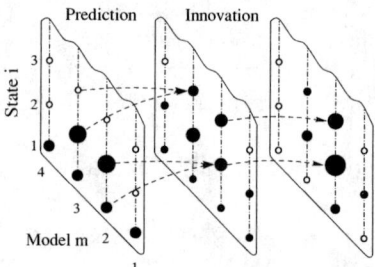

Figure 3. Recognition of Temporal Structures using Semi Conditional Density Propagation. The plane on the left shows the discrete model state distribution $p(q_{t-1}^m = i|O_{t-1})$ at time $t - 1$. The plane in the middle shows the predicted model state distribution $p(q_t^m = j|O_{t-1})$ at time t. New state probabilities for a state j in model m are predicted according to equation 5 by multiplying all states i in the previous plane with the state transition probabilities a_{ij}^m for the given model m. The plane on the right shows the new model state distribution $p(q_t^m = j|O_t)$ at time t after multiplying all states with the corresponding observation probability $b_j^m(o_t)$ equation 6.

review Condensation and introduce a new discrete representation.

Classical Condensation is defined by two equations:
Prediction:

$$p(s_t|O_{t-1}) = \int_{s_{t-1}} p(s_t|s_{t-1}) p(s_{t-1}|O_{t-1}) \qquad (1)$$

where s_t describes a continuous state at time t that will only depend on its previous state $p(s_t|s_{t-1}) = p(s_t|S_{t-1})$ at time $(t - 1)$ and is independent of its former history $S_{t-1} = (s_1, s_2, \ldots, s_{t-1})$. Whilst $p(s_t|O_{t-1})$ is the "predicted" state distribution for the accumulated observation history $O_{t-1} = (o_1, o_2, \ldots, o_{t-1})$ at time $(t - 1)$ and $p(s_t|s_{t-1})$ is the state transition probability from a state s_{t-1} to s_t.

Innovation:

$$p(s_t|O_t) = k_t \, p(o_t|s_t) \, p(s_t|O_{t-1}) \qquad (2)$$

where $p(s_t|O_t)$ is the conditional state distribution at time t for the accumulated observation history O_{t-1}, $p(s_t|O_{t-1})$ is the "predicted" prior and $p(o_t|s_t)$ the conditional observation density and k_t a normalisation factor.

3.5 Semi Continuous States

Replacing the continuous state variable s_t with a discrete state q_t^m for model m, allows us to replace the state transition probabilities $p(s_t|s_{t-1})$ with

$$p(s_t|s_{t-1}) = p(q_t^m = j|q_{t-1}^m = i) = a_{ij}^m \qquad (3)$$

the probability for model m to be in state $q_t = i$ at time t and q_j at $t + 1$, and the conditional observation density $p(o_t|s_t)$ with

$$p(o_t|s_t) = p(o_t|q_t^m = j) = b_j^m(o_t) \qquad (4)$$

the probability for model m to observe o_t in state $q_t = j$. This allows us to rewrite the above *Prediction*

$$p(q_t^m = j|O_{t-1}) = \sum_i a_{ij}^m p(q_{t-1}^m = i|O_{t-1}) \qquad (5)$$

and the *Innovation*

$$p(q_t^m = j|O_t) = k_t b_j^m(o_t) p(q_t^m = j|O_{t-1}) \qquad (6)$$

The above equations resemble condensation except that we are working with discrete states and therefore have no need to approximate a continuous state density distribution using a discrete number of samples. Therefore factored sampling becomes obsolete and computational costs are reduced.

3.6 Recognition

Recognition can now be performed as illustrated in Figure 3. Initially we have neither information on the current observation sequence nor have prior knowledge on the individual model probabilities. Therefore we assume all models m have an equal probability and initialise the probability $p(q_o^m = 1|O_{-1})$ of being in the first state given the at the moment not yet existing observation history O_{-1}

to 1 divided by the number of models. Next we perform a prediction according to Equation (5) and predict new state probabilities $p(q_1^m = j|O_{-1})$ for all states j and all models m. The predicted states are updated and multiplied with $p(o_0|q_0^m = j)$, the current probability of observing o_0 in state j according to the Innovation in Equation (6). The last two steps are repeated until either all states in a particular model m are traversed and the probability $p(q_t^m = LastStateOfModelM|O_t)$ of being in the last state of model m is above a preset threshold or we reach the last observation in the current data segment. In the former case we have recognised a known gesture and can rebuild the model, for the later case we have to add a new model to our model set.

4 Experiments

To evaluate our approach, we recorded a participant performing 8 gestures in arbitrary order. A time limit was set to 10 minutes and each gesture had to be performed 10 times. The gestures where recorded using a Polhemus tracker, an electromagnetic tracking device that is able to determine the 3D position of a small sensor relative to the center of a transmitter. Two sensors were attached to the subject's body, one to the head, used for reference and one to the right hand. The 3D positions of both sensors were recorded with 5 frames per second and the relative difference projected onto a virtual image plane, thus creating an 2 dimensional observation vector containing the x and y position of the subjects hand relative to the head. The complete sequence was segmented into 25 clusters using k-means, analysed for potential rest positions and used to generate a codebook for the Semi Continuous Markov Models. Extracted clusters and overlaid trajectories are depicted in Figure 4.

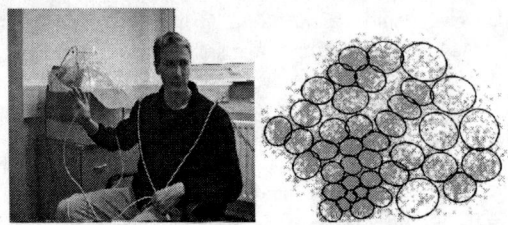

Figure 4. Experimental Setup: The Polhemus tracker with sensors attached to the right hand and head (left). The recorded test sequence projected onto a virtual image plane, divided into 25 clusters using k-means (right).

The first column of Figure 5 , Figure 6 and Figure 7 show example trajectories for each gesture. Figure 5 shows examples for gestures that have been segmented into two phases. They are, *pointing left, pointing right* and *"please sit down"*. Figure 6 shows examples for gestures that have

been segmented into three phases. They are *"there was a big explosion"*, *"he bent a tree"* and *"the shape of a woman"*. Figure 7 shows examples for gestures that have been segmented into four phases. They are, *waving high* and *waving low*. Columns to the right show the corresponding and automatically acquired atomic models with highlighted mixture components. For example, the first row in Figure 5 shows three images. The first on the left shows the trajectory of the hand moving through gesture space. The second image shows the acquired model for the movement into gesture space and the third the acquired model for the movement out of gesture space and back into rest position. A total of 57 atomic components was acquired. However 36 were based on less than 3 repetitive sequences and only the 21 most frequently seen are depicted. The small numbers inside each cluster indicates model states, sharing the same mixture component and the number in the top left of each model shows the number of repetitive sequences, the model is based on and therefore can be seen as measure for the performance and recognition rate. This number should be 10 in an ideal case, except for the second and third phase of the *waving high* and *waving low* gesture. The hand moves twice from the left to the right, therefore it should be 20. The second column of the *pointing right* gesture (Figure 5, row 2) and the second column of the *" there was a big explosion"* gesture (Figure 6, row 1) however have an occurrence count of 17, which indicates that those atomic models are shared by both gestures. Respectively for the first atomic model in *" he bent a tree"* and *"the shape of a woman"* gesture. (Figure 6, row 2 and 3).

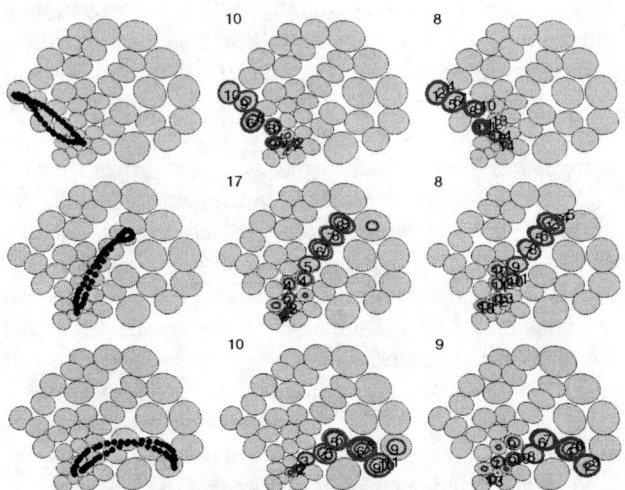

Figure 5. Gestures and corresponding models composed of two atomic components. From top to bottom, *pointing left* , *pointing right* and *"please sit down"*.

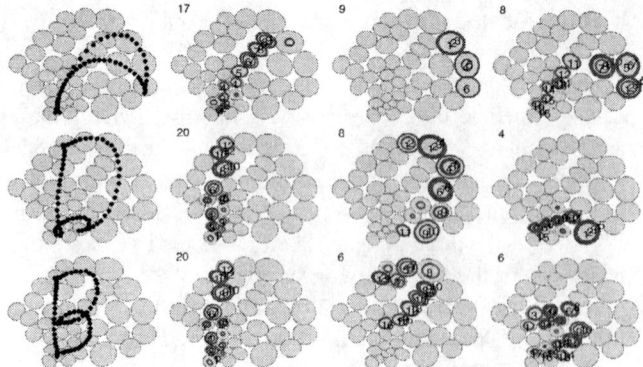

Figure 6. Gestures and corresponding models composed of three atomic components. From top to bottom, *"there was a big explosion"* , *"he bent a tree"* and *"the shape of a woman"* .

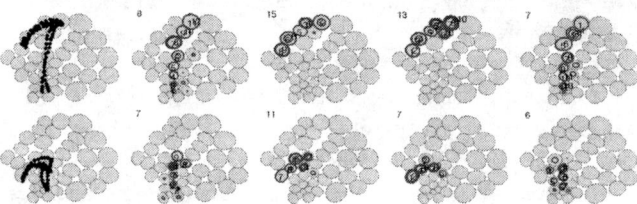

Figure 7. Gestures and corresponding models composed of four atomic components. From top to bottom *waving high* and *waving low*.

5. Conclusion and Future Work

We proposed an approach to automatically segment a continuous input stream into atomic components, using only contextual information derived from the observation sequence itself and to incrementally extract a set of atomic components, for the automatic model acquisition of natural gestures. The method is based on the assumption that gestures can be viewed as a repetitive sequence of atomic components, similar to phonemes in speech, governed by a high level structure controlling the temporal sequence. So far higher level knowledge is only represented as structure containing the temporal sequences of atomic components enclosed, and the number of occurrences in between two rest positions. Therefore future implementations will model the high level structure in a more appropriate way. We will extend the approach to continuously re-cluster the gesture space and thus will be able to build models at a finer scale. We will also investigate methods to fuse prematurely segmented atomic components to reduce their number.

References

[1] H. Asada and M. Brady. The curvature Primal Sketch. *Technical Report.*,MIT AI memo 758, 1984

[2] M. Black and A. Jepson. Recognizing temporal trajectories using the condensation algorithm. In *IEEE Conf. on Face & Gesture Recognition*, pages 16–21, Nara, Japan, 1998.

[3] A. Bobick and A. Wilson. A state-based technique for the summarization and recognition of gesture. In *Proc. Int. Conf. Comp. Vis.*, 1995.

[4] S.A. Engel, and J.M. Rubin. Detecting visual motion boundaries. In *Proc. Workshop on Motion: Representation and Analysis*, Charleston, SC. pp. 107-111, 1986

[5] A. Galata, N. Johnson and D. Hogg. Learning Behaviour Models of Human Activities. In *BMVC*, Nottingham, UK, 1999. pp. 12–22

[6] X.D. Huang and M.A. Jack. Semi Continuous Hidden Markov Models for Speech Signals. In *Computer Speech and Language.*,vol 3, no 3, 1989

[7] M. Isard and A. Blake. Contour tracking by stochastic propagation of conditional density. In *ECCV*, pages 343–357, Cambridge, UK, April 1996.

[8] N. Johnson. *Learning object behaviour models.* PhD thesis, University of Leeds, England, September 1998.

[9] S. McKenna and S. Gong. Gesture recognition for visually mediated interaction using probabilistic event trajectories. In *BMVC*, pages 498–508, Southampton, UK, 1998.

[10] D. McNeill. Hand and Mind: What Gestures Reveal About Thought. *Univ. of Chicago*, 1992.

[11] L. Rabiner and B. Juang. *Fundamentals of Speech Recognition.* Prentice Hall, New Jersey, USA, 1993.

[12] J. Schlenzig, E. Hunter and R. Jain. Recursive Identification of Gesture Inputs Using Hidden Markov Models. In *Proc. Second Annual Conf. Applications of Computer Vision*,pp. 187-194, Dec. 1994.

[13] T. Starner and A. Pentland. Visual recognition of American Sign Language using Hidden Markov Models. In *Proc. of the Intl. Workshop on Automatic Face- and Gesture-Recognition*, Zurich, 1995.

[14] C. Vogler and D. Metaxas. Towards scalability in ASL recognition: Breaking down signs into phonemes. *Gesture Workshop*, Gif sur Yvette, France, 1999.

[15] M. Walter, A. Psarrou and S. Gong. Learning Prior and Observation Augmented Density Models for Behaviour Recognition. In *BMVC*, 1999. pp. 23–32

[16] A. Wilson, A. Bobick and J. Cassell. Temporal Classification of Natural Gesture and Application to Video Coding. In *Computer Vision and Pattern Recognition*, 1997.

[17] A. Wilson and A. Bobick. Realtime Online Adaptive Gesture Recognition. In *Proceedings of the International Workshop on Recognition, Analysis, and Tracking of Faces and Gestures in Real-Time Systems*,Corfu, Greece, Sept. pp. 26-27, 1999.

[18] J.L. Yamato, J. Ohya and K. Ishii Recognizing human action in time-sequntial images using hidden Markov model. *Proc. Conf. on Computer Vision and Pattern Recognition*, Champaign, IL 1992, pp. 379-385.

Motion Estimation & Recognition II

Individual Recognition from Periodic Activity Using Hidden Markov Models

Qiang He and Chris Debrunner
Colorado School of Mines
Division of Engineering
{qhe, cdebrunn}@mines.edu

Abstract

We present a method for recognizing individuals from their walking and running gait. The method is based on Hu moments of the motion segmentation in each frame. Periodicity is detected in such a sequence of feature vectors by minimizing the sum of squared differences, and the individual is recognized from the feature vector sequence using hidden Markov models. Comparisons are made to earlier periodicity detection approaches and to earlier individual recognition approaches. Experiments show the successful recognition of individuals (and their gait) in frontoparallel sequences.

1. Introduction

In automated visual surveillance systems, recognition of humans and their activities is generally the most important task. Video retrieval systems also can also use such capabilities to expand the range of queries they can handle. Two forms of human recognition can be useful in these contexts: the determination that an object is from the class of humans (which we call human recognition), and determination that an object is a particular individual from this class (which we call individual recognition). In this paper we focus on the latter problem, but we expect the approach to be equally applicable to the former. In addition, the approach can successfully distinguish periodic activities.

Human action can be either periodic or non-periodic, and in our approach we processes these two cases differently. In this paper, we describe a technique for recognizing individuals from periodic motions, specifically walking and running. Periodicity is an important component of the information in an activity, so we emphasize periodicity detection in our approach.

Figure 1 shows three frames from an example sequence processed by our approach.

Our approach consists of the following steps. We first segment the image sequence based on motion, and we compute the Hu moments of segmented motion regions. We then match Hu moments over time to determine the degree of periodicity and the period of the motion. Then hidden Markov models (HMMs) are used to recognize the individuals from sequences of quantized Hu moment vectors. The periodicity information is used to compute the size of codebook, to compute the number of states and to distinguish the two kinds of activities (walking and running).

Figure 1   Three frames from an example of a walking sequence.

2. Previous work

The works most related to ours are Yamato *et al.* [1]* and Little and Boyd [9]. Yamato *et al.* made use of hidden Markov models to recognize the human actions based on low-resolution image intensity patterns in each frame. These patterns were passed to a vector quantizer, and the resulting symbol sequence was recognized using a HMM. Their method did not consider the periodicity information,

---

* Many other methods make use of HMMs for gesture or action recognition [2-8], but Yamato *et al.* use simple features more similar to the Hu moments we use.

and they also have no systematic method for determining the parameters of the vector quantization.

Our approach differs in that it explicitly models and uses the periodicity, and in that it uses features based only on the moving portion of the image (background pixels are not used in the feature computation).

Davis and Bobick [10] used Hu moments of motion history images (MHI) and motion energy images (MEI) computed over a short time period from the foreground regions in each frame. These images (and their Hu moments) are distinctive for a variety of actions, and can therefore be used for action or gesture recognition. Davis and Bobick recognize the extracted Hu moment vectors by matching them to trained Gausssian distributions in the features space using the Mahalanobis distance. As in our approach, these features are only invariant to viewpoint to the extent that the Hu moments provide invariance to rotation, translation, reflection, and scale. This work focused on non-periodic actions and hence did not consider the periodicity information.

Little and Boyd [9] compute a set of moment features of the optic flow. The time sequence of each feature is then analyzed for periodicity and the phase difference between features is computed. These phase differences are the quantities used for recognizing individuals. They demonstrate high recognition rates in a large data set collected under very controlled circumstances. Our approach recognizes motion segmentations instead of optic flow, which we feel is a more stable set of features. Our method achieves recognition rates similar to Little and Boyd over a smaller data set, but one with larger variations in the collection environment.

Detection and estimation of image sequence periodicity has been more thoroughly explored in previous work than individual recognition. Generally speaking, there are two types of methods for periodicity detection: those based on the Fourier transform and those implemented in the time domain. Typically, a stochastic signal can be decomposed into the deterministic (periodic) component and the non-deterministic (random) component after Fourier transformation [11]. In frequency domain, the deterministic component corresponds to the harmonic peaks and the non-deterministic component corresponds to the smooth part of the spectrum. Therefore, the energy of the spectral harmonic peaks is a good measure of the periodicity. In Polona's method [12], a periodicity measure based on 1-D Fourier transforms along the temporal dimension is computed. The periodicity measure is defined for each pixel as the normalized difference of the sum of the power spectrum values at the highest amplitude frequency and its multiples, and the sum of the power spectrum values at the frequencies halfway between. If the ratio is close to one, the motion is periodic, otherwise it is not periodic. The periodicity for the entire image is defined as the maximal value of the average value of each special periodicity measure. Tsai makes use of the Fourier transform of the autocorrelation of the curvature of a spatio-temporal curve to study cyclic motion [13], and uses a median filter to estimate the background. Liu and Picard [11] measure the amount of periodicity in a tracked object pixel using the *temporal harmonic energy ratio*, which is the ratio of the energy in the spectral peaks to the total energy in the spectrum.

The second type of method implements the periodicity analysis in time domain. Little and Boyd [9] used maximum entropy spectrum estimation to find the period. Allmen and Dyer [14] used curvature as a low-level description of motion, and its scale-space as a representation. An advantage of curvature scale-space is that it is possible to detect cycles at any scale. Our method computes a minimum in the feature vector difference, and is reliable across our entire dataset.

3. Algorithm description

The training phase and the recognition phase of our approach make use of the following processing steps: motion segmentation, feature extraction, periodicity detection, feature quantization, and HMM recognition. In the training phase, motion segmentation is performed on each frame and features are extracted. The periodicity is determined from the features, and the number of frames in the period is used as the number of states in the HMM. One codebook and one HMM is trained for each gait of each individual. The training sequences are used to generate the codebook for the vector quantization [15]. The training sequences are converted into symbol sequences using this codebook, and the HMM is trained from theses symbol sequences.

In the recognition phase, motion segmentation is performed on each frame and features are extracted. The periodicity is determined and is used to determine the gait. For each trained HMM of the selected gait, the features are quantized into symbol sequences, and the probability that the HMM produced the symbol sequence is computed (see the forward algorithm in [16]). The output of the recognizer is the individual and gait of the HMM with the maximum probability.

The following subsections describe the processing steps in more detail.

  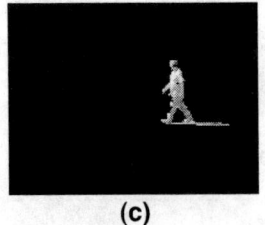 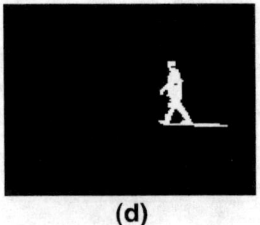

| (a) | (b) | (c) | (d) |

Figure 2   The steps of the motion segmentation process. (a) shows the original image, (b) shows the pixels above the change threshold, (c) shows the image of (b) after median filtering, and (d) shows the resulting binarized segmentation. Note that the shadow is detected too, which causes failures in our activity recognition approach (as well as in others).

3.1. Motion segmentation

In order to study the cyclic human action, we need to segment the objects from the background. This is accomplished by robustly estimating the statistics (mean) of the background and segmenting any pixels that do not fit the statistics. Median filtering along the temporal direction is performed for each pixel to estimate the mean of its background value. Because of wind and unstable light sources, small blobs always appear. We make use of spatial median filtering in each segmented image to remove the small blobs. The Figure 2 is an example of typical segmentation results.

3.2. Feature extraction

Hu moments as described in [17] and [18] remain constant under translation, rotation, reflection, and scaling, so they are good descriptions for image segmentations computed using the algorithm described in the previous section. We use the Hu moments of the segmented images as the image feature vectors.

3.3. Periodicity detection

The periodicity detection is based on the similarity for Hu moments,

$$N_t = \arg\min_{i \in S} |H_t - H_{t+i}|^2,$$

where N_t is the number of image frames in one cycle starting at frame t. H_t is the Hu moment vector at time t, S is the search range. The time interval between two contiguous image frames is known, allowing us to compute the periodicity from N_t. However, the actual periodicity is double the computed periodicity because of the symmetry of walking and running in the side-views

that we use. This method is simple, and experiments show that it is more exact and more reliable than the Fourier transformation methods we tested. At the same time, tracking of complex features is unnecessary. As is common in periodicity estimation, we use the average value of the periodicities computed over several cycles as the overall periodicity of each activity. Figure 3 shows the two matching patterns detected using this method.

| Frame 14 | Frame 31 |

Figure 3   Examples of the matching frames selected by the Hu moment matching method of periodicity estimation.

3.4. Feature quantization

The Hu moment feature vectors are quantized to provide a discrete set of symbols as input to the HMM. We generate the codebook by averaging the corresponding feature vectors from several successive cycles of the training sequence. The codebook size we use is the number of frames in a period. Given the codebook, input feature vectors can be converted to symbols by finding the closest codebook vector [15]. The symbols corresponding to the closest codebook vectors are passed as input to the HMMs.

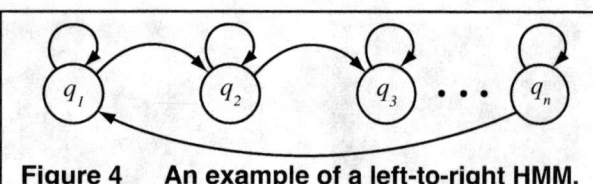

Figure 4   An example of a left-to-right HMM.

3.5. Hidden Markov models

It has been shown that hidden Markov models are successful tools for modeling and classifying dynamic behaviors. For example, HMMs have been widely and successfully applied to speech recognition. There are many similarities between speech recognition and human action recognition. A human being's voice and human action are both dynamic behaviors. Each person has characteristic speech and action patterns. If the characteristic features of a person's speech or actions are extracted, they can be recognized by a HMM.

In order to model human gait, we make use of left-to-right discrete HMM (Figure 4.), where one state of the HMM represents one gait state. We select half the number of frames in one cycle as the number of states in the HMM. For our left-to-right discrete HMM, there are two possible transitions in one HMM state, one to the same state and one to the next state.

The HMM approach to individual recognition is actually a classification method, that is:

$$M = \arg\max_{j=1,2,\ldots,C} P(M_j \mid S)$$

where S is the input symbol sequence coming from the vector quantization of the motion feature vectors, M_j is the jth HMM, and C is the number of HMMs.

HMM parameter estimation makes use of an Expectation Maximization (EM) approach called the Baum-Welch algorithm. The HMMs are trained on the symbol sequences generated from the vector quantized Hu moment vectors computed from the motion segmentations of the training sequences.

4. Experiments

The experiments use image sequences of three individuals (who we will refer to as A, B, C) in both walking and running gaits. We assume that there exists one constant viewpoint and the human activities are frontoparallel to the camera and with constant speed within a particular sequence. The image sequences were collected using a Cannon Optura miniDV camera, were captured to a computer at 30 frames per second with 640 by 480 resolution, and were converted to grayscale before

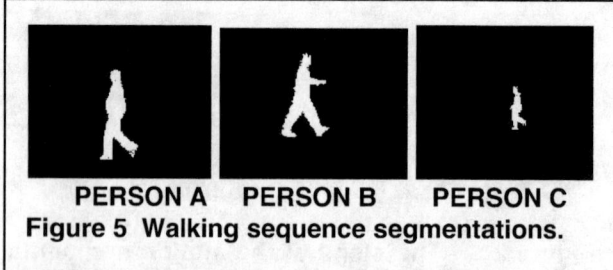

Figure 5 Walking sequence segmentations.

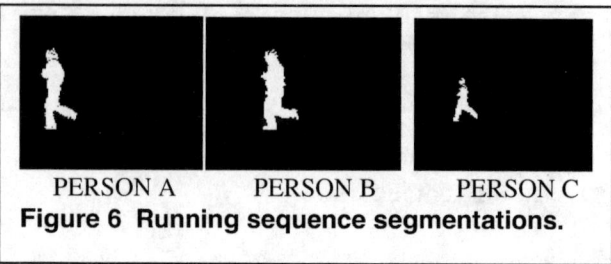

Figure 6 Running sequence segmentations.

motion segmentation. There are ten sequences for each individual in each activity. Each sequence contained many cycles of the activity. Two of each of these ten are used to train a HMM and eight are used to test the HMM. For an input observation sequence, we set a periodicity threshold of 24 frame/cycle to distinguish walking and running. The ten sequences for a particular individual and activity had significant variation in backgrounds, walking and running speeds, directions of motion (left or right), distances to the camera, and attire of the individuals.

Unlike the experiments of Little and Boyd [9], some of the sequences were collected in lighting that produced shadows. To eliminate this effect, these sequences were manually edited to remove the shadows. An alternative would be to detect shadows based on chrominance as suggested in [19].

Table 1 and Table 3 show periodicity computed for the walking and running sequences, respectively. Table 2 and Table 4 show a typical log-likelihood output for example walking and running sequences, respectively. Table 5 shows the recognition rate computed over all sequences. Figure 5 shows the typical segmentations from the walking sequences. Figure 6 shows typical segmentations from the running sequences.

Table 1   Walking periodicity.

	Computed Period	Actual Period
Person A Walk	16 frame	15 frame
Person B Walk	13 frame	13 frame
Person C Walk	15 frame	15 frame

We have found that due to the spatial median filtering in the motion segmentation step, the scale of the subject (which depends on a variety of factors including distance to the subject and focal length) has an influence on the computed Hu moment feature values and hence on recognition accuracy. Because of the median filtering, some of the finer features of the subject silhouette are eliminated when the subject is at a smaller scale. Because our data for person C included several sequences taken at a greater distance than typical in our data, this effect is reflected in the lower recognition rate for person C .

Table 2   An example of the computed HMM log probabilities for three walk test sequences. Rows correspond to trained HMMs, and colunms correspond to test sequences.

	Person A	Person B	Person C
Person A Walk	-74.9798	-470.282	-433.428
Person B Walk	-501.518	-107.909	-181.961
Person C Walk	-77.2505	-144.244	-124.003

Table 3   Running periodicity.

	Computed Period	Actual Period
Person A Run	9 frames	10 frames
Person B Run	12 frames	12 frames
Person C Run	10 frames	11 frames

Table 4   An example of the computed HMM log probabilities for three run test sequences. Rows correspond to trained HMMs, and colunms correspond to test sequences.

	Person A	Person B	Person C
Person A Run	-55.1951	-61.4325	-132.271
Person B Run	-87.2595	-47.644	-118.383
Person C Run	-149.381	-112.923	-51.0995

Table 5   Recognition rate for individuals.

	Recognition Rate (%)
Person A	100% (16/16)
Person B	93.75% (15/16)
Person C	87.5% (14/16)

5. Conclusion

In this paper, a simple and reliable method is given for analyzing the periodicity of human activity and for recognizing individuals. Our approach is robust over variations in backgrounds, walking and running speeds, direction of motion (left or right), attire of the individuals, and, to some degree, distance to the camera. One limitation common to our approach and most previous approaches is that the motion must be frontoparallel. We expect that our approach would perform similarly in other fixed viewpoints if the training is also performed in the same viewpoint. An approach with a fixed viewpoint can find applications in surveillance (e.g., monitoring a hallway or an entrance where everyone passes in one of two orientations).

While the simplicity of this approach is appealing, it is unclear whether it will extend well to variable viewpoints or to larger numbers of individuals. One approach we will explore to expand the range of viewpoints is the use of affine invariant moments [18] rather than Hu moments. These features are invariant to more general transformations of the data, but the shorter feature vector may contain less information that is unique to individuals. The Hu moment features may also not be a rich enough feature set for distinguishing larger groups of individuals (although using time sequences certainly provides a high dimensional feature space). We will also explore richer feature sets such as joint angles, although these features may be difficult to extract reliably from video.

6. References

1. Yamato, J., J. Ohya, and K. Ishii. *Recognizing human action in time-sequential images using hidden Markov model.* in *Computer Vision and Pattern Recognition.* 1992. p. 379-385.
2. Brand, M., N. Oliver, and A. Pentland. *Coupled hidden Markov models for complex action recognition.* in *Computer Vision and Pattern Recognition.* 1997. San Jaun, PR.
3. Wilson, A.D. and A.F. Bobick, *Parametric Hidden Markov Models for Gesture Recognition.* IEEE Transactions on Pattern Analysis and Machine Intelligence, 1999. **21**(9): p. 884-900.
4. Starner, T. and A. Pentland. *Visual Recognition of American Sign Language Using Hidden Markov Models.* in *International Conference on Automatic Face and Gesture Recognition.* 1995. Zurich, Switzerland.
5. Thrun, S., J.C. Langford, and D. Fox. *Monte Carlo hidden Markov models: Learning non-parametric models of partially observable stochastic processes.* in *ICML-99.* 1999.
6. Pentland, A. and A. Liu, *Modeling and Prediction of Human Behavior,* . 1995, M.I.T. Media Lab Perceptual Computing.
7. Bregler, C. *Learning and Recognizing Human Dynamics in Video Sequences.* in *Computer Vision and Pattern Recognition.* 1997. San Jaun, PR.
8. Wilson, A.D. and A.F. Bobick. *Nonlinear PHMMs for the Interpretation of Parameterized Gesture.* in *Computer Vision and Pattern Recognition.* 1998. Santa Barbara, CA.
9. Little, J.J. and J.E. Boyd, *Recognizing People by Their Gait: The Shape of Motion.* Videre: Journal of Computer Vision Research, 1998. **1**(2): p. 2-32.
10. Davis, J.W. and A.F. Bobick. *The Representation and Recognition of Action Using Temporal Templates.* in *Computer Vision and Pattern Recognition.* 1997. San Jaun, PR. p. 928-934.

11. Liu, F. and R.W. Picard. *Finding periodicity in space and time*. in *International Conference on Computer Vision*. 1998. Bombay, India.

12. Polana, R. and R. Nelson. *Detecting Activities*. in *Computer Vision and Pattern Recognition*. 1993. New York, NY. p. 2-7.

13. Tsai, P.-S. and M. Shah, *Cyclic Motion Detection*, . 1993, University of Central Florida.

14. Allmen, M.C. and C.R. Dyer. *Cyclic motion detection using spatiotemporal surfaces and curves*. in *International Conference on Pattern Recognition*. 1990. p. 365-370.

15. Gray, R.M., *Vector Quantization*. IEEE ASSP Magazine, 1984(4): p. 4-29.

16. Rabiner, L.R. and B.H. Juang, *An introduction to hidden Markov models*. IEEE ASSP Magazine, 1986(January 1986): p. 4-16.

17. Hu, M.K., *Visual pattern recognition by moment invariants*. IRE Transactions Information Theory, 1962. **8**(2): p. 179-187.

18. Sonka, M., V. Hlavac, and R. Boyle, *Image processing, analysis, and machine vision*. 2 ed. 1999: PWS Publishing.

19. Wren, C.R., *et al.*, *Pfinder: Real-time tracking of the human body*. IEEE Transactions on Pattern Analysis and Machine Intelligence, 1997. **19**(7): p. 780-785.

On the Improvement of Anthropometry and Pose Estimation from a Single Uncalibrated Image

Carlos Barron and Ioannis A. Kakadiaris
Department of Computer Science
University of Houston
4800 Calhoun, Houston, TX 77204-3475
{cbarron, ioannisk} @uh.edu

Abstract

Recently, we developed a technique that allows semi-automatic estimation of anthropometry and pose from a single image. However, estimation was limited to a class of images for which adequate number of human body segments were almost parallel to the image plane. In this paper, we present a generalization of that estimation algorithm that exploits pairwise geometric relationships of body segments to allow estimation from a broader class of images. In addition, we refine our search space by constructing a fully populated discrete hyper-ellipsoid of Stick Human Body Models (SMs) in order to capture the variance of the statistical anthropometric information. As a result, a better initial estimate can be computed by our algorithm and thus the number of iterations needed during minimization are reduced by tenfold. We present our results over a variety of images to demonstrate the broad coverage of our algorithm.

1. Introduction

Most of the research in the area of human motion analysis is directed to tracking humans in (monocular or multi-camera) image sequences [12], due to the wide range of applications that could benefit from estimating and analyzing human motion. These applications include surveillance, performance measurement for athletes and patients with disabilities, motion capture for entertainment applications, and vision-based user interfaces. However, the important problem of estimating an individual's anthropometry and pose from a single uncalibrated image has received considerably less attention although it constitutes the first step in many human tracking (from monocular images) algorithms.

Recently, we developed a technique which allows semi-automatic estimation of anthropometry (up to a scale parameter) and pose from a single image [2, 3]. The user initially selects a set of image points that constitute the projection of selected landmarks. Using this information, along with a priori statistical information about the human body, a set of plausible segment length estimates are generated. The third step produces a set of plausible poses based on joint limit constraints using a geometric method. In the fourth step, pose and anthropometric measurements are obtained by minimizing an appropriate cost function subject to the associated constraints. The novelty of that approach was the use of anthropometric statistics to constrain the estimation process that allows the simultaneous estimation of both anthropometry and pose. However, estimation was limited to a class of images for which adequate number of human body segments were almost parallel to the image plane. For example, that method could handle images like the one depicted in Figure 1(a), but not images like the one depicted in Figure 1(b). In this paper, we present a generalization of that reconstruction algorithm to allow estimation from a broader class of images including the one depicted in Figure 1(b). In particular, first we explore the projective properties of segments that have pairwise similar orientation with respect to the camera. Second, we extend our technique for exploiting prior statistical anthropometric information to allows us to obtain a better initial estimate of the human body model. As a result, the number of iterations needed by our algorithm to converge is reduced ten times. Also note that our algorithm can signal the absence of adequate information for producing a reliable estimate.

The remainder of this paper describes our enhancements in more detail. In Section 2, we review prior work in the area, and in Section 3 we provide our analysis and the proposed enhancements. In particular, in Section 3.2 we explore how to exploit additional geometric relationships of the human subject's body segments by considering their foreshortening in the image, and in Section 3.3 we describe our refinements of the prior statistical models to further constrain the selection of initial estimates for the minimization

(a)   (b)

Figure 1. (a) Instance of an image that can be handled by the algorithm described in [[2,3]], (b) Instance of an image that could not be handled using [[2,3]] but it can be handled using the algorithm described in this paper.

process. Finally, in Section 4 we illustrate results from our system.

2. Prior Work

One of the challenges in model-based human tracking algorithms is the initialization of the model in the first frame of the image sequence. Tracking and posture estimation methods have been presented that use either one [15, 22, 5, 6, 4, 9, 14, 18, 23], or multiple cameras [1, 10, 13, 7, 8, 11]. However, in most of the existing tracking approaches the user specifies an approximate position and posture from the human model at the first frame of the image sequence [6, 13, 19]. In contrast, Bregler and Malik [5] for the initialization step of their human tracking method, they minimize a cost function over position, angles and body dimensions. However, no information is provided about the accuracy and repeatability of their method, nor for what class of postures and human body dimensions does the method work. Recently, Taylor [21] presented a method for recovering information about the configuration of articulated objects from a single image. The similarities with our work are that both methods assume a scaled orthographic projection and that they both use geometrical information as constraints. The difference is that our work combines the geometrical constraints with the prior statistical anthropometric information to drive the estimation process. Thus, our method begins and ends with a plausible human model. Rosales and Sclaroff [20] presented a method for pose estimation based on selecting the most likely pose given the learned probability distribution and the visual features' similarity between hypothesis and input. For our algorithm, there is no need for a learning phase. Moeslund and Granum [16] represented

the human model in phase space spanned by its different degrees of freedom and they used an analysis-by-synthesis approach to match the phase space model with real images for the case of a human arm model. They currently examine what would be the impact of a significantly increased size of the phase space (as in the case of estimating the pose of a full human body model) in the efficiency of their algorithm. The contribution of our paper is a systematic study and an improved technique that takes into consideration statistical anthropometric information to constrain the estimation process.

3. Analysis

The problem of anthropometry and pose estimation from a single image can be formulated as follows: *Given a set of points in an image that correspond to the projection of landmark points of a human subject, estimate both the anthropometric measurements (up to a scale) of the subject and his/her pose that best match the observed image.* In the following it is assumed that the selected landmarks at the image are the result of a scaled orthographic projection of three-dimensional landmark points of a human subject.

3.1. Stick Human Body Model

For the purposes of this research, we have developed a generic Stick Human Body Model, whose complete description can be found in [2, 3]. Briefly, a Stick Model is a tree $(s, \mathcal{S}, \mathcal{A})$, where $\mathcal{S}$ is a set of sites/landmarks and $\mathcal{A}$ is a collection of edges (segments) with endpoints in $\mathcal{S}$, and $s \in \mathcal{S}$ is the root. In our case, $\mathcal{A}=\{$ HD, RY, LY, NK, UT, RC, LC, RUA, LUA, RLA, LLA, RHD, LHD, LT, RHP, LHP, RUL, LUL, RLL, LLL, RF, LF$\}$ as enumerated in Figure 2(b), and the set of landmarks consists of a set of joints $\mathcal{J}=\{$at, sp, la, lc, le, lh, lk, ls, lw, ra, rc, re, rh, rk, rs, rw, wt$\}$ (information about the SM's joints is provided in Table 1), and other landmarks M=\{ry (right eye), ly (left eye), rhd (base of the right middle finger), lhd (base of the left middle finger), rf (tip of the right foot), lf (tip of the left foot)\} $(\mathcal{S} = \mathcal{J} \cup \mathcal{M})$. The default data for the joints and the anthropometric measurements are extracted from [17].

3.2. Exploiting Geometric Relationships

In this section, we will examine the foreshortening of the body segments in the image, under the assumption of scaled orthographic projection. Let $\mathbf{C} = [X_C, Y_C, Z_C]^\top$ be the origin of the camera (see Fig. 3) and let's assume that the image plane is located at Z_{IM} on the z-axis of the camera. As known, under scaled orthographic projection the point $\mathbf{P_1} = [X_1, Y_1, Z_1]^\top$ (see Fig. 3) projects to the point $\mathbf{p_1} = [x_1, y_1, z_1]^\top = [X_C + \lambda_1(X_1 - X_C), Y_C + \lambda_1(Y_1 - Y_C),$

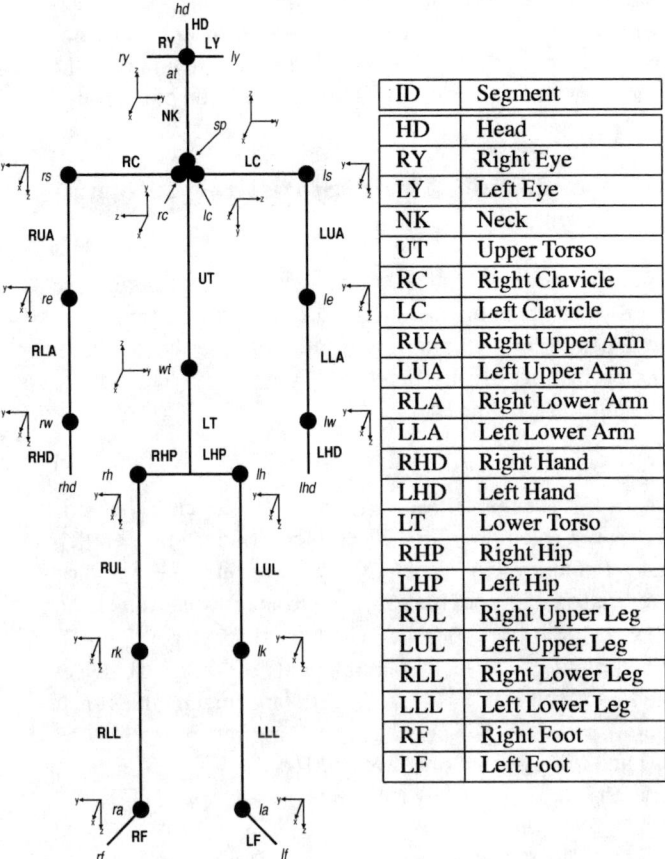

ID	Segment
HD	Head
RY	Right Eye
LY	Left Eye
NK	Neck
UT	Upper Torso
RC	Right Clavicle
LC	Left Clavicle
RUA	Right Upper Arm
LUA	Left Upper Arm
RLA	Right Lower Arm
LLA	Left Lower Arm
RHD	Right Hand
LHD	Left Hand
LT	Lower Torso
RHP	Right Hip
LHP	Left Hip
RUL	Right Upper Leg
LUL	Left Upper Leg
RLL	Right Lower Leg
LLL	Left Lower Leg
RF	Right Foot
LF	Left Foot

Figure 2. (a) Stick Human Body Model (SM) and its associated coordinate systems, (b) Names of the SM's segments (adapted from [2,3])

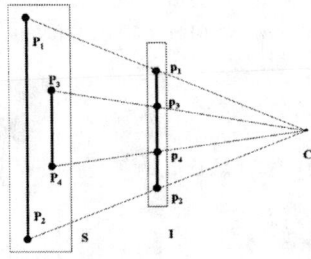

Figure 3. Notation pertaining to the scaled orthographic projection camera model assumed in this paper.

ID	Joint	From	To	DOF
at	atlanto occipital	NK	HD	Tz*Rz*Ry*Rx
sp	solar plexus	UT	NK	Tz*Ry*Rz*x
la	left ankle	LLL	LF	Tx*Rz*Rx*Ry
lc	left clavicle	UT	LC	Tz*Rx*Ry
le	left elbow	LUA	LLA	Tz*Ry
lh	left hip	LT	LUL	Tz*Rz*Rx*Ry
lk	left knee	LUL	LLL	Tz*R-y
ls	left shoulder	LC	LUA	Tz*Rz*Rx*Ry
lw	left wrist	LLA	LHD	Tz*Ry*Rx*Rz
ra	right ankle	RLL	RF	Tx*R-z*R-x*Ry
rc	right clavicle	UT	RC	Tz*R-x*Ry
re	right elbow	RUA	RLA	Tz*Ry
rh	right hip	LT	RUL	Tz*R-z*R-x*Ry
rk	right knee	RUL	RLL	Tz*R-y
rs	right shoulder	RC	RUA	Tz*R-z*R-x*Ry
rw	right wrist	RLA	RHD	Tz*Ry*R-x*R-z
wt	waist	LT	UT	Tz*Ry*Rz*Rx

Table 1. Information related to the joints of the Stick Model [2,3].

$Z_{IM}]^\top$, where $\lambda_1 = \frac{Z_{IM} - Z_C}{Z_1 - Z_C}$. Similarly, the point $\mathbf{P_2} = [X_2, Y_2, Z_2]^\top$ projects to the point $\mathbf{p_2} = [x_2, y_2, z_2]^\top = [X_C + \lambda_2(X_2 - X_C), Y_C + \lambda_2(Y_2 - Y_C), Z_{IM}]^\top$, where $\lambda_2 = \frac{Z_{IM} - Z_C}{Z_2 - Z_C}$. If we assume that this point lies on the same plane (normal to the camera z-axis) with point $\mathbf{P_1}$, then $\lambda_1 = \lambda_2$. Thus, for any point on the line $\overline{\mathbf{P_1P_2}}$, its projection is given by the equation: $[x, y]^\top = \lambda_1 \mathbf{S}[X, Y, Z]^\top$, where $\mathbf{S} = \begin{bmatrix} 1 & 0 & 0 \\ 0 & 1 & 0 \end{bmatrix}$. Similarly, for any point on the line $\overline{\mathbf{P_3P_4}}$, its projection is given by the equation: $[x, y]^\top = \lambda_3 \mathbf{S}[X, Y, Z]^\top$, where $\lambda_3 = \frac{Z_{IM} - Z_C}{Z_3 - Z_C} = \frac{Z_{IM} - Z_C}{Z_4 - Z_C} = \lambda_4$. If α_z is a real number such that

$$(1 + \alpha_z) = \frac{Z_1 - Z_C}{Z_3 - Z_C} , \qquad (1)$$

then $Z_3 - Z_C = \frac{Z_1 - Z_C}{1 + \alpha_z}$ and $\lambda_3 = \lambda_1(1 + \alpha_z)$. Therefore, the scaled orthographic projection for the points of $\overline{P_3P_4}$ is given by: $[x, y]^\top = \lambda_1(1 + \alpha_z)\mathbf{S}[X, Y, Z]^\top$. Let $L_{12} = ||\mathbf{P_2} - \mathbf{P_1}||$ and $l_{12} = ||\mathbf{p_2} - \mathbf{p_1}||$. Then,

$$l_{12} = ((x_2 - x_1)^2 + (y_2 - y_1)^2)^{\frac{1}{2}} =$$
$$= \lambda_1((X_2 - X_1)^2 + (Y_2 - Y_1)^2)^{\frac{1}{2}} = \lambda_1 L_{12}.$$

Also,
we can obtain that $l_{34} = \lambda_3 L_{34}$, where $L_{34} = ||\mathbf{P_4} - \mathbf{P_3}||$ and $l_{34} = ||\mathbf{p_4} - \mathbf{p_3}||$. Using the relation $\lambda_3 = \lambda_1(1 + \alpha_z)$, we can obtain that $l_{34} = \lambda_1(1 + \alpha_z)L_{34}$. Finally, the ratio between l_{12} and l_{34} is given by: $\frac{l_{12}}{l_{34}} = \frac{\lambda_1 L_{12}}{\lambda_1(1 + \alpha_z)L_{34}} = \frac{L_{12}}{(1 + \alpha_z)L_{34}}$ that implies the following relation:

$$\frac{L_{12}}{L_{34}} = (1 + \alpha_z)\frac{l_{12}}{l_{34}} , \qquad (2)$$

which suggests Proposition 1.

Proposition 1 *For segments that lie in planes almost parallel to the image plane, the ratio of segment lengths in 3D and the ratio of their corresponding projected lengths are similar if and only if α_z is very small.*

Proof. The result is obtained from Eq. (2). ∎

Our previous work [2, 3] was based on the proposition above. In this work, we examine the relationship of segment lengths that lie in different planes using the distance between planes. For example, let's examine the ratio of $L_{13} = \|\mathbf{P_1} - \mathbf{P_3}\|$ and $L_{24} = \|\mathbf{P_2} - \mathbf{P_4}\|$, and the ratio of $l_{13} = \|\mathbf{p_1} - \mathbf{p_3}\|$ and $l_{24} = \|\mathbf{p_2} - \mathbf{p_4}\|$. Since,

$$
\begin{aligned}
l_{13} &= [(x_3 - x_1)^2 + (y_3 - y_1)^2]^{\frac{1}{2}} \\
&= |\lambda_1| \left\{ [(1 + \alpha_z)X_3 - X_1]^2 + [(1 + \alpha_z)Y_3 - Y_1]^2 \right\}^{\frac{1}{2}} \\
&= |\lambda_1| \left\{ L_{13}^2 - (Z_3 - Z_1)^2 + 2\alpha_z[X_3(X_3 - X_1) + \right. \\
&\quad \left. Y_3(Y_3 - Y_1)] + \alpha_z^2(X_3^2 + Y_3^2) \right\}^{\frac{1}{2}},
\end{aligned}
$$

then

$$
\begin{aligned}
\lambda_1^2 L_{13}^2 &= l_{13}^2 + \lambda_1^2 \{ (Z_3 - Z_1)^2 - \\
&\quad 2\alpha_z[X_3(X_3 - X_1) + Y_3(Y_3 - Y_1)] - \\
&\quad \alpha_z^2(X_3^2 + Y_3^2) \}.
\end{aligned}
$$

Using Eq. (1), we obtain $L_{13} = \frac{l_{13}}{\lambda_1}(1 + \epsilon_z)$, where

$$
\begin{aligned}
1 + \epsilon_z &= \{1 + \frac{\lambda_1^2}{l_{13}^2}\{\alpha_z^2[(Z_3 - Z_C)^2 - (X_3^2 + Y_3^2)] - \\
&\quad 2\alpha_z[X_3(X_3 - X_1) + Y_3(Y_3 - Y_1)]\}\}^{\frac{1}{2}}.
\end{aligned}
$$

Similar arguments hold for L_{24} and l_{24}. Thus,

$$
\frac{L_{13}}{L_{24}} = \frac{l_{13}(1 + \epsilon_{13})}{l_{24}(1 + \epsilon_{24})}, \tag{3}
$$

where ϵ_{ij} depends on the position of points $\mathbf{P_i}$, $\mathbf{P_j}$ and $\mathbf{C}$.

Proposition 2 *Let's assume that two segments have the following two properties: a) their endpoints lie on planes perpendicular to the camera's Z axis, and b) their orientation w.r.t. the camera is almost similar. Then the ratio of lengths of these two segments is similar to the ratio of lengths of their corresponding projected segments.*

Proof. As per our hypotheses ϵ_{13} and ϵ_{24} must have similar values and thus the conclusion follows from Eq. (3). ∎

In summary, the following two observations hold: a) for segments that are almost parallel to the image plane α_z takes a small value (as per Eq. (2)), and b) for segments that have orientation almost parallel to each other the corresponding factors $1 + \epsilon_z$ for these segments must have similar value (as per Eq. (3)). Based on these observations, during the initialization step of our algorithm the user marks the segments that have orientation almost parallel to the image plane (same as in [2, 3]), and in addition it marks any segments that have similar orientation with respect to the camera. These projected segments are employed to compute the ratios that will be used as input for the selection of the initial Stick Model.

3.3. Exploiting Prior Statistical Information

Using prior statistical anthropometric information we build our cadre family as a multivariate representation of the extremes of the population distribution as described in [2, 3]. The probability density function of the multivariate normal distribution is defined by:

$$
f(\mathbf{x}) = ((2\pi)^k |\Sigma|)^{-\frac{1}{2}} \exp^{[-\frac{1}{2}(\mathbf{x} - \mathbf{m})^{\top} \Sigma^{-1} (\mathbf{x} - \mathbf{m})]}, \tag{4}
$$

where k is the number of dimensions (in our case the variables are the lengths of the 22 segments of our Stick Model, $\mathbf{x}$ is a random vector, $\mathbf{m}$ and Σ are the mean and the covariance matrix of the population, and we use the quadratic form $Q(\mathbf{x}) = (\mathbf{x} - \mathbf{m})^{\top} \Sigma (\mathbf{x} - \mathbf{m})$ whose shape depends on Σ. We compute the principal components of Σ and we select the first seven $\mathbf{e}_i (i = 1, \dots, 7)$ with large eigenvalues such that $\lambda_1 > \lambda_2 > \cdots > \lambda_7$. The following notation relates the SM and the eigenvectors of Σ : $\text{SM}(\alpha) = \sum_{i=1}^{7} \alpha_i \mathbf{e}_i + \mathbf{m}$, where $\boldsymbol{\alpha} = [\alpha_1, \dots, \alpha_7]^{\top}$ such that:

$$
\sum_{i=1}^{7} \alpha_i^2 \leq 1. \tag{5}
$$

That is, we are using linear combinations of the eigenvectors constrained by Eq. (5) to span on the hyper-ellipsoid of Σ. In particular, we build a family of q different models ($\text{SM}(\boldsymbol{\alpha}_q)$), and we compute ratios of lengths $r_{k,q}$ as follows:

$$
r_{k,q} = \begin{cases} \frac{l_{m,q}}{l_{n,q}} & \text{if } \mu(l_n) > \mu(l_m) \\ \frac{l_{n,q}}{l_{m,q}} & \text{otherwise,} \end{cases} \tag{6}
$$

where $\mu(l_n)$ is the length of the segment l_n for the SM $\mathbf{m}$. Furthermore, we compute the covariance coefficients for these ratios as follows:

$$
v_{kj} = \mu\left[(r_{k,q} - \mu(r_k))(r_{j,q} - \mu(r_j)) \right].
$$

We select an initial SM that satisfies the following equation:

$$
q^* = \arg \min_{\boldsymbol{\alpha}_q \in \Delta} \sum_{k \in \mathcal{K}} (r_{k,q} - s_k)(\sum_{j \in \mathcal{K}} v_{kj}(r_{j,q} - s_j)), \tag{8}
$$

where $r_{k,q}, r_{j,q}$ are ratios obtained from the SMs in our cadre family and s_k, s_j are the corresponding ratios computed from the input data as follows:

$$
s. = \begin{cases} \frac{l_m}{l_n} & \text{if } \mu(l_n) > \mu(l_m) \\ \frac{l_n}{l_m} & \text{otherwise} \end{cases}
$$

56

Our objective is to find a finite subset of $\Delta = \{\boldsymbol{\alpha} = [\alpha_1 \ldots \alpha_7]^{\top} \in \mathbf{R}^7 | \sum_{i=1}^{7} \alpha_i^2 \leq 1\}$ in order to build our cadre family $\text{SM}(\boldsymbol{\alpha}_q)$ that will provide a good initial estimate $\text{SM}(\boldsymbol{\alpha}_{q^*})$. The complete algorithm for the SM selection is described in [3].

Intuitively, we build all the linear combinations of the eigenvectors with at least one non zero coefficient under the constraint that all non zero coefficients must be equal, and we minimize Eq. (8) using a discrete cadre family. Note that in Eq. (8) we can use $\sum_{i=1}^{7} \alpha_i^2 = 1$ instead of $\sum_{i=1}^{7} \alpha_i^2 \leq 1$. The following proposition justifies this modification.

Proposition 3 *If* $0 < \sum_{i=1}^{7} \alpha_i^2 < 1$ *then* $\exists \, \boldsymbol{\alpha}' = [\alpha_1', \ldots, \alpha_7']^{\top}$ *such that* $\sum_{i=1}^{7} \alpha_i'^2 = 1$.

Proof. Indeed, $\boldsymbol{\alpha}' = \dfrac{1}{\sqrt{\sum_{i=1}^{7} \alpha_i^2}} [\alpha_1, \ldots, \alpha_7]^{\top}$. $\blacksquare$

Therefore, we can restrict Δ to $\Delta^D = \{\boldsymbol{\alpha} = [\alpha_1, \ldots, \alpha_7]^{\top} \in \mathbf{R}^7 | \sum_{i=1}^{7} \alpha_i^2 = 1 \vee \boldsymbol{\alpha} = 0\}$. To analyze the extended set of selected models $\{\text{SM}(\boldsymbol{\alpha}) | \boldsymbol{\alpha} \in \Delta^D\}$, we introduce a metric space using the quadratic form Q, since Δ^D depends on the matrix $\boldsymbol{\Sigma}$. Let $\hat{\mathbf{e}}_i = \frac{\mathbf{e}_i}{\|\mathbf{e}_i\|}$, $i = 1, \ldots, 7$ where $\|\cdot\|$ is the Euclidean norm of $\mathbf{R}^{22}$. The following proposition presents of the matrix $\boldsymbol{\Sigma}$ and an explanation of the bounds of our cadre family.

Proposition 4 *For every* $(\boldsymbol{\alpha}, \boldsymbol{\beta}) \in \Delta^D$ *and for every unit vector* $\hat{\mathbf{e}}_i$ *the following properties hold:*

1. *The inner product induced by* $\boldsymbol{\Sigma}$, $(\boldsymbol{\alpha}, \boldsymbol{\beta})_{\boldsymbol{\Sigma}} = \boldsymbol{\alpha}^{\top} \boldsymbol{\Sigma} \boldsymbol{\beta}$, $\forall \boldsymbol{\alpha}, \boldsymbol{\beta} \in \mathbf{R}^{22}$ *is commutative.*

2. $(\hat{\mathbf{e}}_i, \hat{\mathbf{e}}_i)_{\boldsymbol{\Sigma}} = \lambda_i \neq 0$, *and* $(\hat{\mathbf{e}}_i, \hat{\mathbf{e}}_j)_{\boldsymbol{\Sigma}} = 0, i \neq j$ *(thus* $\{\hat{\mathbf{e}}_i\}$ *is an orthogonal set).*

3. $\lambda_7 \leq \left(\sum_{i=1}^{7} \alpha_i \hat{\mathbf{e}}_i\right)^{\top} \boldsymbol{\Sigma} \left(\sum_{i=1}^{7} \alpha_i \hat{\mathbf{e}}_i\right) \leq \lambda_1$.

The proposition 3 explains why we restrict $\boldsymbol{\alpha}$ by the constraint $\sum_{i=1}^{7} \alpha_i^2 = 1$. The following equality holds: $\text{SM}(\boldsymbol{\alpha}) = \sum_{i=1}^{7} \alpha_i \mathbf{e}_i + \mathbf{m} = \sum_{i=1}^{7} \alpha_i \|\mathbf{e}_i\| \hat{\mathbf{e}}_i + \mathbf{m}$. Thus, $Q(\text{SM}(\boldsymbol{\alpha})) = \left(\sum_{i=1}^{7} \alpha_i \|\mathbf{e}_i\| \hat{\mathbf{e}}_i\right)^{\top} \boldsymbol{\Sigma} \left(\sum_{i=1}^{7} \alpha_i \|\mathbf{e}_i\| \hat{\mathbf{e}}_i\right)$, and by the previous proposition we obtain the following inequality:

$$\lambda_7 \min_{i=1,\ldots,7} \left\{\|\mathbf{e}_i\|^2\right\} \leq Q\left(\sum_{i=1}^{7} \alpha_i \|\mathbf{e}_i\| \hat{\mathbf{e}}_i\right) \leq \lambda_1 \max_{i=1,\ldots,7} \left\{\|\mathbf{e}_i\|^2\right\}.$$

This expression provides a theoretical bound for the extension of hyper-ellipsoid of $Q(\text{SM}(\boldsymbol{\alpha}))$ with center in $\mathbf{m}$.

In the following, we describe our new full discretization $\Delta^{\overline{D}}$ of Δ^D. Let $\mathcal{I} = \{1, 2, \ldots, 7\}$ and $\mathcal{K} \subseteq \mathcal{I}$. We select $\boldsymbol{\alpha} \in \mathbf{R}^7$ such that $\sum_{i \in \mathcal{I}} \alpha_i^2 = 1$, $\alpha_j = 0, \forall j \notin \mathcal{I}(\mathcal{K})$, and $(\alpha_i = \alpha_j) \forall i, j \in \mathcal{K}$. For example, from this formulation, we can use $\alpha_1 = \alpha_2 = \ldots = \alpha_7 = \pm\frac{1}{\sqrt{7}}$, or $\alpha_1 = 0$, and $\alpha_2 = \ldots = \alpha_7 = \pm\frac{1}{\sqrt{6}}$, etc. This formulation extents our previous approach and the number of SMs that we employ is $\sum_{k=0}^{7} \binom{7}{k} 2^k = 2187$. The coefficients and the number of elements for each SM are given in Table 2.

Coefficient	Linear Combinations
$\pm\frac{1}{\sqrt{7}}$	$\binom{7}{7} 2^7 = 128$
$\pm\frac{1}{\sqrt{6}}$	$\binom{7}{6} 2^6 = 448$
$\pm\frac{1}{\sqrt{5}}$	$\binom{7}{5} 2^5 = 672$
$\pm\frac{1}{\sqrt{4}}$	$\binom{7}{4} 2^4 = 560$
$\pm\frac{1}{\sqrt{3}}$	$\binom{7}{3} 2^3 = 280$
$\pm\frac{1}{\sqrt{2}}$	$\binom{7}{2} 2^2 = 84$
± 1	$\binom{7}{1} 2^1 = 14$
0	1

Table 2. Factor and its number of linear combinations.

The distribution of the $\text{SM}(\boldsymbol{\alpha})$, $\boldsymbol{\alpha} \in \Delta^{\overline{D}}$ can estimated in terms of $\boldsymbol{\Sigma}$ and $\{\hat{\mathbf{e}}_i\}$.

Proposition 5 *Given* $\boldsymbol{\alpha}, \boldsymbol{\beta} \in \Delta^{\overline{D}}$ *then*

$$2\lambda_7 \left(1 - \sum_{i=1}^{7} \alpha_i \beta_i\right) \leq \left\|\sum_{i=1}^{7} (\alpha_i - \beta_i) \hat{\mathbf{e}}_i\right\|_{\boldsymbol{\Sigma}} \leq 2\lambda_1 \left(1 - \sum_{i=1}^{7} \alpha_i \beta_i\right), \text{and}$$

$$2\lambda_1 \left(1 - \sum_{i=1}^{7} \alpha_i \beta_i\right) = \sum_{i=1}^{7} (\alpha_i - \beta_i)^2 \lambda_i$$

Proof. The result follows from $\left\|\sum_{i=1}^{7} (\alpha_i - \beta_i) \hat{\mathbf{e}}_i\right\|_{\boldsymbol{\Sigma}} = \sum_{i=1}^{7} (\alpha_i - \beta_i)^2 \lambda_i$. $\blacksquare$

However, this approach provides a large number of possible combinations for the initial anthropometric estimates. In particular, if $\boldsymbol{\alpha} \in \Delta^{\overline{D}}$, then

$$\text{SM}(\boldsymbol{\alpha}) = \sum_{i=1}^{7} \alpha_i \mathbf{e}_i + \mathbf{m} = $$
$$= \left[l_{1m} + \sum_{i=1}^{7} \alpha_i \delta l_{1i}, \ldots, l_{22m} + \sum_{i=1}^{7} \alpha_i \delta l_{22i}\right]^{\top}$$

where $\mathbf{m} = [l_{1m}, l_{2m}, \ldots, l_{22m}]^{\top}$ is the average human model, and $\mathbf{e}_i = (\delta l_{1i}, \delta l_{2i}, \ldots, \delta l_{22i})$, $i = 1, \ldots, 7$ are the eigenvectors of Σ. For example, for a particular α that corresponds to the index q the ratios $r_{k,q}$ are given by the equation (6) as follows:

$$r_{k,q} = \frac{l_{m,q} + \sum_{i=1}^{7} \alpha_{i,q} \delta l_{mi}}{l_{n,q} + \sum_{i=1}^{7} \alpha_{i,q} \delta l_{ni}}.$$

The terms $\sum_{i=1}^{7} \alpha_{i,q} \delta l_{mi}$ and $\sum_{i=1}^{7} \alpha_{i,q} \delta l_{ni}$ are used to compute the ratio and they depend on the selection of $\alpha_{i,q}$.

The benefit from using a larger cadre family than one used in our previous approach is the improvement of the initial estimation using few ratios only. It enables finding a solution for the images that depict humans in complicated postures and pruning the user selected ratios in order to compute a good initial SM which in turn improves the accuracy of our minimization process.

3.4. Algorithm

Our technique for simultaneously estimating the anthropometry and the pose from a single uncalibrated image has the following steps:

Algorithm: Anthropometry and Pose Estimation
Step 1: Selection of projected landmarks and ratio computation.
Step 2: Ratio pruning and choice of initial Stick Model.
Step 3: Initial estimates for pose.
Step 4: Iterative minimization over lengths and angles.

In this work, we have modified the first and second step of our algorithm presented in [2, 3] as described in the following paragraphs.

Selection of projected landmarks and ratio computation:
We have extended the user-interface developed for [2, 3] to allow a user not only to select the projection of the visible landmarks of a subject's body, but also to define the ratios to be used for the initial estimates using the properties described in Section 3.2. Each ratio is treated independently, since the properties described in Section 3.2 are local and are valid for at least two segments. Note that the selected segments must be in parallel planes or in a similar orientation.

Ratio pruning and choice of initial Stick Model: Our basic assumption is that there is a number of segments that have the properties described in Section 3.2. The user defines the input ratios from the segments that have one of these properties. In this step, our algorithm compares the selected ratios with the range of our cadre family and selects the ones that lie inside the range. For the case that at least one selected ratio lies within the range of our cadre family, the algorithm selects a SM using the technique described in [3]. This step weighs the ratios using the Mahalanobis

distance in order to select a model that closely matches to the input ratios obtained from the image. Otherwise, the user is being informed that the selected ratios lie outside of the cadre family's range (that means that our algorithm cannot handle this image), and s/he is being asked to select additional ratios, if possible.

4. Experimental Results

We have performed a number of experiments on synthetic and real data to assess the accuracy, limitations, and advantages of our approach. In the first experiment, we applied our technique to an image created using the virtual human modeling tool EAI Jack®. Figures 4(a,b) depict the reconstructed 3D model from two views. Tables 3 and 4 contain statistical information related to the accuracy of the estimation process.

In the second experiment, we applied our technique to a real image from the subject *Vanessa* whose anthropometric dimensions were manually measured. Figure 4(c) depicts the selected points, and Figure 4(d) depicts the reconstructed model from a novel view. Table 5 captures the percentage errors (PE) of the estimation process. We observe that the estimation of anthropometric information is within 1% of the anthropometric dimensions of the subject, clearly an improvement of our previous method whose accuracy was within 3.2%. In general, we have performed numerous other experiments with a variety of subjects whose anthropometric dimensions are known with similar, very encouraging results.

In the third experiment, we applied our algorithm to a variety of images from a variety of application domains, where anthropometric information about the subjects was not available, Figures 4(e-g) depict the input images along with the selected points and the reconstructed model from various viewpoints.

	UT+LT LF	UT+LT LLL	UT+LT LLA	UT+LT LUA	LUL LLL	LLL LLA
Actual	2.53	1.31	2.92	2.48	0.84	2.23
Estimated	2.52	1.31	2.93	2.49	0.84	2.23
PE %	0.40	0.00	0.41	0.56	0.05	0.11

Table 3. Accuracy of the length estimates for the synthetic experiment.

5. Conclusions

In this paper, we have investigated the problem of recovering the anthropometry (up to a scale parameter) and the pose of human figure from a single image. Specifically, we have presented our extensions to our previously developed method that allows estimation from a broader class of

Joint	Actual Values	Estimated	PE %
at	$(0.0°,0.0°,0.0°)$	$(0.00°,0.00°,0.00°)$	0.00
sp	$(-44.0°,0.0°,0.0°)$	$(-44.05°,0.08°,0.0°)$	0.21
la	$(10.0°,-20.00°,10.0°)$	$(10.02°,-19.90°,10.0°)$	0.42
lc	$(0.0°,0.0°)$	$(0.0°,0.0°)$	0.00
le	$(90.0°)$	$(90.11°)$	0.12
lh	$(0.0°,15.0°,10.0°)$	$(0.0°,15.40°,9.96°)$	2.23
lk	$(15.0°)$	$(15.15°)$	1.00
ls	$(44.0°,45.0°,145.0°)$	$(43.38°,44.68°,145.31°)$	0.48
lw	$(0.0°,0.0°,0.0°)$	$(0.0°,0.0°,0.0°)$	0.00
ra	$(0.0°,0.0°,0.0°)$	$(0.0°,0.0°,0.0°)$	0.00
rc	$(0.0°,0.0°)$	$(0.0°,0.0°)$	0.00
re	$(110.0°)$	$(110.02°)$	0.02
rh	$(0.0°,15.0°,50.0°)$	$(-0.18°,15.04°,49.87°)$	0.43
rk	$(55.0°)$	$(54.94°)$	0.11
rs	$(70.0°,50.0°,20.0°)$	$(70.24°,50.05°,19.91°)$	0.30
rw	$(0.0°,0.0°,0.0°)$	$(0.0°,0.0°,0.0°)$	0.00
wt	$(25.0°,0.0°,0.0°)$	$(25.15°,-0.12°,0.01°)$	0.77

Table 4. Accuracy of the pose estimates for the synthetic experiment.

	$\frac{UT+LT}{LF}$	$\frac{UT+LT}{LLL}$	$\frac{UT+LT}{LLA}$	$\frac{UT+LT}{LUA}$	$\frac{UL}{LLL}$	$\frac{LL}{LLA}$
Actual	2.43	1.16	2.59	2.23	0.87	2.23
Estimates [2,3]	2.35	1.15	2.55	2.17	0.87	2.22
PE %	3.29	0.86	1.54	2.69	0.00	0.45
New Estimates	2.43	1.15	2.60	2.24	0.87	2.23
PE %	0.00	0.86	0.39	0.22	0.00	0.00

Table 5. Accuracy of the length estimates for the subject *Vanessa*.

images. Unlike other approaches, the novelty of the proposed methods lies in exploiting effectively prior statistical anthropometric information to constrain the estimation process and allow the simultaneous estimation of both anthropometry and pose. Experimental results on a variety of images indicate that our method produces accurate results for a broad class of images.

6. Acknowledgments

We wish to thank Honda R&D Americas for their financial support.

References

[1] A. Azarbayejani, C. Wren, and A. Pentland. Real-time 3-D tracking of the human body. In *Proceedings of the IMAGE'COM 96*, Bordeaux, France, May 1996.

[2] C. Barrón and I. A. Kakadiaris. Estimating anthropometry and pose from a single image. In *Proceedings of the 2000 IEEE Computer Society Conference on Computer Vision and Pattern Recognition*, pages 669–676, Hilton Head Island, SC, June 13-15 2000.

[3] C. Barrón and I. A. Kakadiaris. Estimating anthropometry and pose from a single image. *Computer Vision and Image Understanding*, 2000.

[4] R. Bowden. Learning statistical models of human motion. In Kakadiaris and Sharma [12], pages 10–17.

[5] C. Bregler and J. Malik. Tracking people with twists and exponential maps. In *Proceedings of the 1998 IEEE Computer Society Conference on Computer Vision and Pattern Recognition*, pages 8–15, Santa Barbara, CA, June 23-25 1998.

[6] T. Cham and J. Rehg. A multiple hypothesis approach to figure tracking. In *Proceedings of the 1999 IEEE Computer Society Conference on Computer Vision and Pattern Recognition*, volume II, pages 239–245, Fort Collins, Colorado, June 23-25 1999.

[7] M. M. Covell, A. Rahimi, and T. J. Darrell. Articulated-pose estimation using brightness- and depth-constancy constraints. In *Proceedings of the 2000 IEEE Computer Society Conference on Computer Vision and Pattern Recognition*, pages 438–445, Hilton Head Island, SC, June 13-15 2000.

[8] Q. Delamarre and O. Faugeras. 3D articulated models and multi-view tracking with silhouettes. In *Proceedings of the 7th International Conference on Computer Vision*, pages 716–721, Kerkyra, Greece, September 20-27 1999. IEEE Computer Society.

[9] J. Deutscher, A. Blake, and I. Reid. Articulated body motion capture by annealed particle filtering. In *Proceedings of the 2000 IEEE Computer Society Conference on Computer Vision and Pattern Recognition*, pages 126–133, Hilton Head Island, SC, June 13-15 2000.

[10] D. M. Gavrila and L. S. Davis. 3-D model-based tracking of humans in action: a multi-view approach. In *Proceedings of the 1996 IEEE Computer Society Conference on Computer Vision and Pattern Recognition*, pages 73–80, San Francisco, CA, June 18-20 1996.

[11] S. Iwasawa, J. Ohya, K. Takahashi, T. Sakaguchi, S. Kawato, K. Ebihara, and S. Morishima. Real-time, 3D estimation of human body postures from trinocular images. In *IEEE International Workshop on Modeling People*, pages 3–10, Corfu, Greece, September 20 1999.

[12] I. Kakadiaris and R. Sharma, editors. *Proceedings of the IEEE Workshop on Human Modeling, Analysis and Synthesis*, Hilton Head Island, SC, June 16 2000. IEEE Computer Society Press, New York, NY.

[13] I. A. Kakadiaris and D. Metaxas. Model-based estimation of 3D human motion with occlusion based on active multi-viewpoint selection. In *Proceedings of the 1996 IEEE Computer Society Conference on Computer Vision and Pattern Recognition*, pages 81–87, San Francisco, CA, June 18-20 1996.

[14] I. A. Kakadiaris and D. Metaxas. Model-based estimation of 3D human motion. *IEEE Transactions on Pattern Analysis and Machine Intelligence*, 2000. To Appear.

[15] H. J. Lee and Z. Chen. Determination of 3D human body postures from a single view. *Computer Vision, Graphics and Image Processing*, 30:148–168, May 1985.

[16] T. B. Moeslund and E. Granum. 3D human pose estimation using 2D-data and an alternative phase space representation. In Kakadiaris and Sharma [12], pages 26–33.

[17] National Aeronautics and Space Administration. Man systems integration standards. Technical report, National Aeronautics and Space Administration, 1987.

[18] D. Ormoneit, H. Sidenbladth, M. Black, T. Hastie, and D. Fleet. Learning and tracking human motion using functional analysis. In Kakadiaris and Sharma [12], pages 2–9.

[19] R. Plankers, P. Fua, and N. D'Apuzzo. Automated body modeling from video sequences. In *IEEE International Workshop on Modeling People*, pages 45–52, Corfu, Greece, September 20 1999.

[20] R. Rosales and S. Sclaroff. Inferring body pose without tracking body parts. In *Proceedings of the 2000 IEEE Computer Society Conference on Computer Vision and Pattern Recognition*, pages 721–727, Hilton Head Island, SC, June 13-15 2000.

[21] C. J. Taylor. Reconstruction of articulated objects from point correspondences. In *Proceedings of the 2000 IEEE Computer Society Conference on Computer Vision and Pattern Recognition*, pages 677–684, Hilton Head Island, SC, June 13-15 2000.

[22] S. Wachter and H.-H. Nagel. Tracking of persons in monocular image sequences. In *Proceedings of IEEE Nonrigid and Articulated Motion Workshop*, pages 2–9, Puerto Rico, June 16 1997.

[23] M. Yamamoto and K. Yagishita. Scene constraints-aided tracking of human body. In *Proceedings of the 2000 IEEE Computer Society Conference on Computer Vision and Pattern Recognition*, pages 151–156, Hilton Head Island, SC, June 13-15 2000.

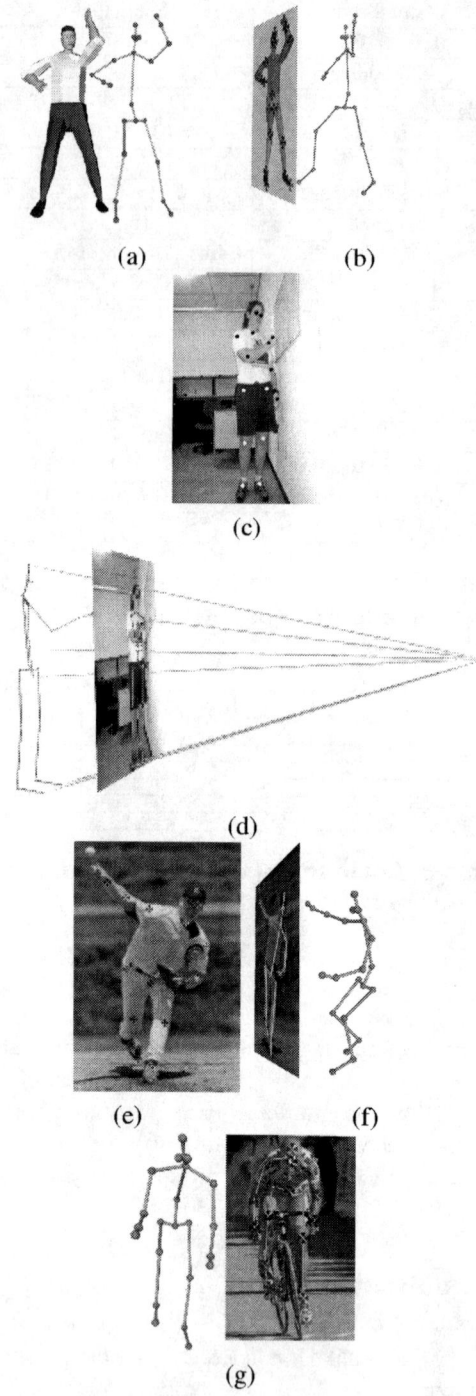

Figure 4. Front (a) and side (b) view of the virtual human along with our estimates. (c) Input image for the subject *Vanessa*, and (d) the reconstructed 3D model. (e) Landmarks selected by a user on this baseball player image, and (f) anthropometry and pose estimates from our algorithm. (g) An example from the domain of cycling.

Human Activity Detection in MPEG Sequences

Burak Ozer[†]   Wayne Wolf[‡]   Ali N. Akansu[†]

†Department of Electrical and Computer Engineering
New Jersey Institute of Technology
New Jersey Center for Multimedia Research
Newark, NJ 07102, USA
ibo8175@oak.njit.edu

‡Department of Electrical Engineering
Princeton University
New Jersey Center for Multimedia Research
Princeton, NJ 08540, USA
wolf@ee.princeton.edu

Abstract

In this paper, we propose a hierarchical method for human detection and activity recognition in MPEG sequences. The algorithm consists of three stages at different resolution levels. The first step is based on the principal component analysis of MPEG motion vectors of macroblocks grouped according to velocity, distance and human body proportions. This step reduces the complexity and amount of processing data. The DC DCT components of luminance and chrominance are the input for the second step, to be matched to activity templates and a human skin template. A more detailed analysis of the uncompressed regions extracted in previous steps is done at the last step via model-based segmentation and graph matching. This hierarchical scheme enables working at different levels, from low complexity to low false rates. It is important and interesting to realize that significant information can be obtained from the compressed domain in order to connect to high level semantics.

1 Introduction

Most human activity recognition techniques are implemented in the uncompressed domain and depend on proper segmentation of the human body. The purpose of our work is to investigate activity recognition in the compressed domain in order to reduce computational complexity and avoid dependency on correct segmentation in an uncompressed image. The algorithm consists of three stages at different resolution levels. The first step is based on the principal component analysis of MPEG motion vectors to match the detected activity with known human activities; namely, walking, running, and kicking. The motion vectors are grouped automatically according to velocity, distance and human body proportions. The algorithm uses DC-DCT coefficients of the luminance and chrominance values when more detailed information is needed. These values are matched to activity templates and a human skin template. The finest details in the sequences are obtained from the uncompressed domain via model based segmentation and graph matching. This hierarchical scheme enables working at different levels, from low complexity to low false rates. Our proposed graph-based representation [1] enables the detection of human presence in the frames, as well as posture recognition by using the DC-DCT coefficients in the compressed and pixel values in the uncompressed domains. The major contribution of the graph matching method is the automatic creation of semantic segments from the combination of low-level edge or region-based segments using model-based segmentation. The generality of the reference model attributes allows the detection of different postures while the conditional rule generation decreases the rate of false alarms.

Since image and video applications are generally represented in the compressed domain, such as JPEG or MPEG, there is a need for image/video manipulation and automatic content extraction in the compressed domain. As stated in [2], for existing compression standards the compressed-domain image/video manipulation techniques can be used to help solve the bandwidth problem. Hence applications without expanding the coded visual content back to the large uncompressed domain would reduce the need of large bandwidth and intensive computing. The use of available information in compressed video and images mostly has been investigated for video indexing, and shot and scene classification. In [3], hierarchical decomposition of a complex video is obtained using scaled DC coefficients in an intra coded DCT compressed video for browsing purposes. The technique combines visual and temporal information to capture the important relations within a scene and between scenes in a video. A general model of a hierarchical scene transition graph is applied for video browsing. In [4], the authors examine the direct reconstruction of DC coefficients from motion compensated P-frames and B-frames

61

of MPEG compressed video. Their analysis and experimental results show that lower cost approximations can be used successfully for various image processing operations, such as shot detection, shot matching and clustering. In [5], an automatic scene classification scheme is proposed for MPEG videos. The scenes are divided into low, medium, and high texture and activity scenes. The bit rates of the I, P and B frames are used in shot texture classification while the percentage of macroblock types are used for shot motion classification. In [6], a metric based on the mean and the variance of MPEG motion vectors is used to classify the scenes according to the activity level.

MPEG motion vectors are used mostly to index videos (low-high activity) and track objects. The object detection in the compressed domain is more restricted since this application requires more detailed information. In [7], an object tracking algorithm is proposed using compressed video only with periodically decoding I-frames. The object to be tracked is initially detected by an accurate but computationally expensive object detector applied to decoded I-frames. In [8], an algorithm to detect human face regions from dequantized DCT coefficients of MPEG video is proposed. The algorithm uses the DC DCT values of chrominance, shape, and energy distributions of the face area. This method is suitable for color images with face regions greater than 48 by 48 pixels (3 by 3 MPEG macroblocks). The authors extend their work in [9] in order to track and summarize faces from compressed video. The previous algorithm is used to detect faces and MPEG motion information is used with the Kalman filter prediction to track faces within each shot. The representative frames are then decoded for pixel domain analysis and browsing.

The data sets for human detection applications in the compressed domain include anchorperson scenes, news stories and interviews, where the faces and the upper-bodies occupy large areas in the image. However, at lower resolution, available motion vectors can be used to detect human activity by comparing it with known human activity patterns. Our work can be divided into three major parts. The first part is activity recognition in the compressed domain based on principal component analysis (PCA) [10]. In Figure 1, the MPEG motion vectors and motion vector groups according to the human body proportion, are displayed. In the second part, DC DCT differences between frames in the compressed domain are matched to activity templates (side-view), obtained from a training set, to distinguish activity periods. The DC coefficients are also used in the graph matching algorithm for human body recognition in the compressed domain, but this method is suitable for images with face regions greater than 3 by 3 macroblocks. Since graph matching performance depends highly on face detection, this is a crucial restriction. In most cases, the resolution of the face area does not satisfy this criteria, which leads us

to implement the graph matching algorithm in the uncompressed domain for the finest analysis of the human body and posture. This paper is organized as follows: Processing in the compressed domain is given in the next section where activity recognition based on principal component analysis is explained. The third section covers the template and graph matching algorithms at higher resolutions. The results are displayed in section 4.

Figure 1. Motion vectors and vector groups.

2 Activity Recognition Using MPEG Motion Vectors

Activity recognition problem can be divided into two subparts: the first one is collecting satisfactory measurements and the second one is developing a recognition algorithm based on these measurements. Most of the related work use activity measurements from uncompressed images after a proper segmentation of human body parts. Our measurements are obtained from MPEG motion vectors of macroblocks. Since the resolution of the motion vectors is one macroblock and there is no direct correspondence with the object parts and their motion, a robust and global model must be used. The corruption of data is another problem in MPEG motion vectors since some blocks can not be tracked during some frames. An overview of research on human motion analysis can be found in [11], [12]. The major problems in the activity recognition is the scale, shift and projection changes between the model and the test data and segmentation dependency. One of the activity modeling methods proposed in [13] is based on first order Markov model descriptions and continuous propagation of observation density distributions. Hidden Markov Models are used to predict the state transitions. In [14], speed and direction components of 2D trajectories are represented by scale-space images that are invariant to Euclidean transforms. A method based on time-frequency analysis is proposed in [15] to detect periodic human motion with self-similar characteristic. The outline of the human body is used to detect the periodical relative limb movement in [16] by a template matching process. In these approaches, for each activity, a separate model is developed in order to compare with the observed activity. These approaches are robust to local transformations but lack a global detailed model to capture the variabilities. Principal component analysis method is one of the global approaches. Our activity recognition model is based on the principal component analysis which has also been used by Yacoob and Black [10] for human activity recognition in uncompressed video sequences. The authors use the motion measurements for segmented human body parts. In our method, we first detect the moving regions and then group the motion vectors automatically by using the ratio of the human body parts. Hence the measurements do not correspond to the actual human body parts but to macroblock groups corresponding to human region. For the classification of moving regions, the neighboring blocks with a velocity greater than a predefined threshold are classified as one moving object. The following subsection covers the principal component analysis.

2.1   Principal Component Analysis

• **Step1:** PCA was successfully used for face recognition. A compact representation of facial appearance is described in [17], where face images are decomposed into weighted sums of basis images using a Karhunen-Loeve expansion. The eigenpicture representation has been used in [18] as eigenfaces for face recognition. PCA is a dimensionality reducing technique, used in pattern recognition. It reduces dimensionality by projecting the motion vectors to a new space spanned by the training data set. For training the system, several walking, running and kicking man sequences which are temporally aligned are used. For these sequences, the object region is extracted by grouping MPEG motion vectors. Then, the object is segmented to three parts (upper body, torso and lower body) according to the human body proportions. The mean of the motion vectors in horizontal and vertical direction is computed for the macroblocks corresponding to each part (6 parameters) for a number of sequences T. A training set of k different examples for each activity forms matrix A of dimensions $6T \times k$. Then the singular value decomposition of the matrix A is computed to get the approximated projection of the exemplar vectors (columns of A) onto the subspace spanned by the $q < k$ basis vectors. Hence activity basis with parameters m are computed.

$$A = U\Sigma V^T \qquad (1)$$

where A is the motion parameter matrix, U represents the principal component directions, Σ includes the singular values, and V^T expands A in principal component directions. To recognize the activities, an unknown sequence, other than test sequences of an activity which can be shifted and scaled in time is compared with the training set. The normalized distance between the coefficients c_i from the observed data set and coefficients of exemplar activities m_i is used to recognize the observed activity that is transformed

Figure 2. Frames from walking, running, and kicking man training sets.

by the temporal translation, scaling and speedup parameters [10]. The Euclidean distance is given as

$$d^2 = \sum_1^q (c_i/\|\mathbf{c}\| - m_i/\|\mathbf{m}\|)^2 \qquad (2)$$

The algorithm is applied for recognition of three activity classes: walking, running, and kicking. 10 training test sequences for each class are obtained from various sources for the side-view. The camera motion is assumed to be zero. In Figure 2, some test frames from the activity training sets are given. The detection of the moving regions and the determination of the activities from the grouped MPEG motion vectors gives a coarse information about the scene.

3   Human Detection and Posture Recognition

The second step of the algorithm uses 8 by 8 block information (DC values) in the frames where human activity has been detected from the motion information. The last step is based on graph matching where nodes correspond to the human body parts in the uncompressed domain. This step is also implemented in the compressed domain where the graph nodes correspond to the segments of DC DCT values.

• **Step2:** The difference of the DC values for 8 by 8 blocks between consecutive frames are computed and the difference image is binarized by thresholding. To train our system, we used several human activity sequences from side-view with the similar camera distance, human motion direction and velocity. In order to find the template for each body position during one activity period, the mean of the moving regions, corresponding to these positions, are calculated. The classification is done by using a basic template matching measure. Note that the mirror image of the template is also used. For every DC-DCT difference frame, the blocks are compared to the activity templates. For scale change invariance, we scaled the moving block regions with different scale parameters and calculated the matching value for each scale factor.

• **Step3:** Note that the sequences of interest include human where in most cases the skin regions consist of one or two 16 by 16 macroblocks. Usually, the skin information from the DCT values of color components can not be used for human detection since the resolution requirement is not

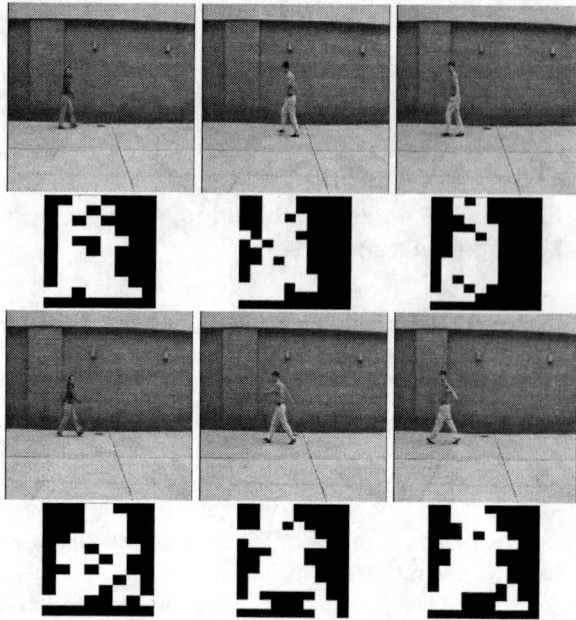

Figure 3. First and third rows: Left column: Walking position from the training set, middle and right columns: Resulting frames with the minimum matching costs. Second and fourth rows: DCT coefficient difference for the corresponding frames.

met. If the skin regions are detected (Figure 4), next step will be the segmentation and implementation of the proposed model based graph matching algorithm on the DC luminance blocks for each frame. As it is mentioned before, finding the head region from the skin color information is a crucial step for the performance of the matching algorithm. For a more detailed investigation, the algorithm is implemented pixel based in the uncompressed domain. Model based graph matching algorithm is summarized in the next subsection.

Figure 4. First and third images: Original frames. Second and fourth images: Marked frames with macroblocks detected as skin regions.

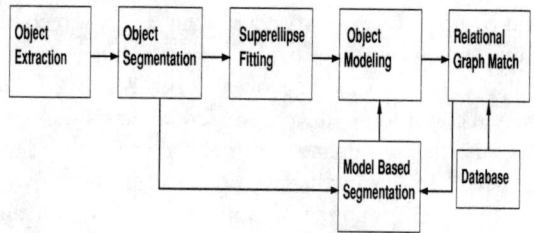

Figure 5. Graph matching algorithm.

3.1 Model Based Graph Matching

This part corresponds to our model based algorithm (Figure 5) [1] for human detection and posture recognition in the uncompressed domain. The major contribution of the proposed method is to create automatically semantic segments from the combination of low level edge or region based segments using model-based segmentation. The segments are modeled by superellipses and Bayes classification is used in the graph matching process. Parametric modeling of image segments helps to overcome the occlusion problem and reduce the effect of the deformations due to the clothing. Graph matching allows to increase the generality of the reference model part attributes for the detection of human with different postures while the conditional rule generation between graph levels decreases the rate of false detection. A brief summary for each algorithm block is given below. In [1], the reader can find a more detailed presentation of the algorithm.

Graph Matching Algorithm

• **Object Segmentation:** Human body consists of several subparts (head, torso, hands, etc.) that can be obtained by segmenting it hierarchically into its smaller unique parts. We use the color image segmentation technique combined with an edge detector algorithm. Skin color model is formed via Farnsworth nonlinear transformation.The segmented image can contain regions corresponding to the background. However, these regions will not match the regions of the template object. The segmented region boundaries can still be in complex forms. The boundaries are first smoothed by a Gaussian smoothing operator. Concave and convex segments (landmarks) that are used for curvature segmentation are determined on the resulting contour.

•**Superellipse Fitting:** Each segmented region is modeled with a superellipse. Even when human body is not occluded by another object, due to the possible positions of non-rigid parts a body part can be occluded in different ways. Therefore, global approximation is preferred.

• **Object Modeling and Similarity Measure:** Description of boundaries for simple object segments using geometric descriptions are computed. These descriptors are classified into two groups: 1) Unary features: a) compactness; b) eccentricity; c) color (hair and skin). 2) Binary features: a) Ratio of areas; b) relative position and orientation; c) the adjacency information between nodes with overlapping boundaries or areas.

• **Relational Graph Matching:** The last part of the algorithm is based on a graph matching approach. Modeling human with relational graphs is widely used in the literature. However, most of them rely on satisfactory segmentation results. The meaningful combinations of classes is used to overcome this problem. In graph representation of human, each level of a branch represents a class for a body part or combination. Each body part and meaningful combinations represent a class. The combination of binary and unary features is represented by a feature vector. For the purpose of determining the class of these feature vectors a piecewise quadratic Bayesian classifier is used. In our case, it is a multiclass and multifeature problem. For the reference model supervised learning is implemented using several test images. The features for each body part are assumed to be Gaussian distributed.

4   Results

To evaluate the system performance for the activity recognition, we used several sequences with different activities. Table 1 displays the resulting normalized distances (Eq. 2) between the activity sets and test sequences. The preliminary results show that MPEG motion vectors corresponding to three human body subregions can be used for detection and recognition of human activity. Each test sequence gives the minimum normalized distance with its corresponding training set. The last sequence is a MPEG car movie. Note that the distances are very high for each activity class. Another restriction for car sequences is that the human body ratio is not suitable for the car mainbody. The performance of the algorithm depends on the temporal duration of the observed activity. The results displayed in the table are given for sequences with two or more activity periods. Different time instants during one period are detected in the second step. Some results are displayed in Figure 3.

Some of the human detection and posture recognition results are also given in this section. Since human body parts are smooth objects the smoothing factor is chosen small ($= 1.25$). Curvature threshold is chosen the same for all the test images ($= 0.55$). In Figure 6, the curvature segmentation result for selected body parts is shown.

The performance of the graph matching algorithm is given for 42 test images for front and side views which are chosen from different sources. In the model file, the adjacency information between parts is given as; head-torso, upper arm-torso, leg-foot, lower arm-hand, etc. For example, there is no adjacency restriction between hand and leg or hand and belly, since hand can be at any position near them. In the model file these combinations are also chosen: arm=upper arm+lower arm, legs=leg1+leg2, lowbody=legs+belly, upbody=torso+belly, armtorso=arm+tor-

	Walking	Running	Kicking
walk1	0.001	0.0587	0.1543
walk2	0.0103	0.0929	0.0615
walk3	0.007	0.02	0.0784
walk4	0.0084	0.1218	0.1627
walk5	0.046	0.1506	0.1651
walk6	0.019	0.1298	0.208
run1	0.26677	0.0954	0.1688
run2	0.2525	0.0143	0.2519
run3	0.7665	0.027	0.1703
kick1	0.298	0.1253	0.0576
kick2	0.1901	0.109	0.0868
car	0.5362	0.4282	0.6922

Table 1.   The normalized Euclidean distances between the activity sets and test sequences.

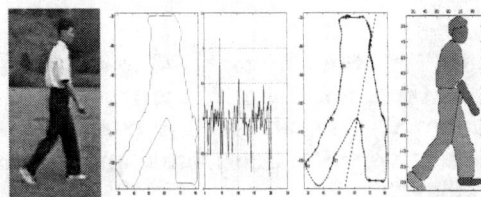

Figure 6.   First: Frame from MPEG7 test sequence. Second: Leg segment. Third: Curvature of the segment ($th_k = 0.55$). Fourth: Curvature segmentation. Fifth: Segmentation result.

so. Results for segmentation and modeling with superellipses are displayed in Figure 7.

After graph matching, the body parts in Figure 7 (first row); face, torso, belly, arm 1, arm 2, leg 1, and leg 2, are correctly classified. Face, torso, and legs (together) are the classified body parts for the second row where the person wears a suit which covers multiple body parts. In the third row, a side-view image is displayed. The correctly classified parts are, face, arm 1, torso, leg 1, leg 2, foot 1, and foot 2. The graph matching algorithm performance is obtained by computing the correct, false, and miss detection of the body parts in the test images. The preliminary results show that % 70.27 of the body parts are correctly and % 18.92 are falsely classified. The remaining % 10.8 is the miss detection. The majority of the falsely classified body parts are hand and foot regions that are generally combined with leg and arm regions. In order to determine the posture of the persons in the still images and video sequences, we use the binary features of the corresponding matched node pairs after the classification. For example, the angle α between the image node matched to torso and image node matched to arm informs how much arms are open. Table 2 displays the results

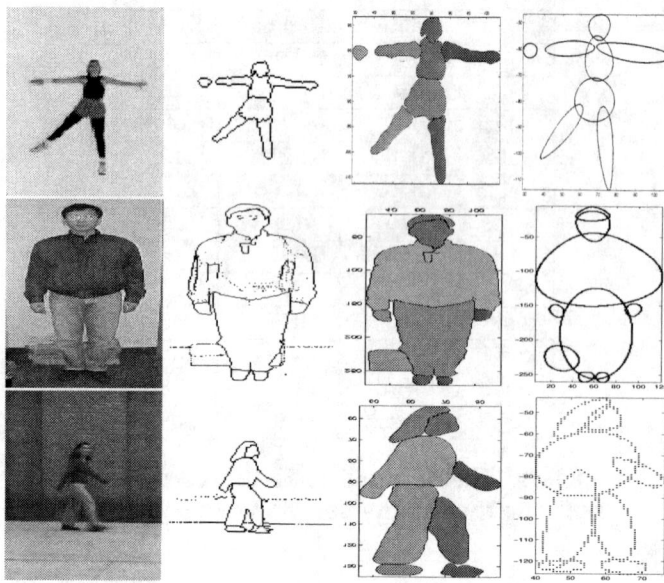

Figure 7. **First column: Original images. Second column: Segmentation results. Third column: Part separation and curvature segmentation results. Fourth column: Fitted superellipses.**

part 1	part2	α
torso	arm 1	79.10
torso	arm 2	75.32
torso	leg 1	39.31
torso	leg 2	2.92

Table 2. **Relative orientation (α) values for Figure 7 (first row).**

for Figure 7 (first row) where both arms are open with an angle of 75-80 degrees, one leg is open with an angle of 40 degrees while other leg is approximately on the same axis as torso. Note that, posture recognition is a direct result of correct classification of the body parts. The segmented body parts can be also used for a more detailed tracking and activity recognition in the uncompressed domain.

5 Conclusions

In this paper, we proposed a hierarchical method for human activity and posture recognition in MPEG sequences for different resolution levels from low complexity to low false rates. The preliminary results indicate that a significant information can be obtained from the compressed domain. Experimental results of principal component analysis show that macroblock motion vectors can be used for activity recognition if the observed activity consists of two or more periods. Detection of human skin regions in the compressed or uncompressed domains increases the performance of the proposed graph matching algorithm.

References

[1] I.B. Ozer, W. Wolf, A.N. Akansu, "Relational Graph Matching for Human Detection and Posture Recognition", SPIE Symposium on Voice, Video, and Data Communications, Nov. 2000, Proceedings of SPIE, Vol. 4210.

[2] S.-F. Chang, J. R. Smith, M. Beigi, and A. B. Benitez, "Visual Information Retrieval from Large Distributed On-line Repositories", Communications of the ACM, Vol. 40, No. 12, 1997, pp. 63-71.

[3] M.M. Yeung, B.L. Yeo, W. Wolf and B. Liu , "Video Browsing using Clustering and Scene Transitions on Compressed Sequences", SPIE Vol. 2417 Multimedia Computing and Networking 1995 , pp. 399-413.

[4] B.L. Yeo and B. Liu , "On the extraction of DC sequence from MPEG Compressed Video", IEEE ICIP, Oct. 1995.

[5] A. M. Dawood and M. Ghanbari, "Scene Content Classification From Mpeg Coded Bit Streams", IEEE Workshop on Multimedia Signal Processing, 1999, pp 253-258.

[6] Kadir A. Peker, A. Aydin Alatan, Ali N. Akansu, "Low-level Motion Activity Features for Semantic Characterization of Video", Proceedings of IEEE International Conference on Multimedia and Expo, 2000.

[7] D. Schonfeld and D. Lelescu, "VORTEX: Video retrieval and tracking from compressed multimedia databases - template matching from MPEG2 video compressed standard", SPIE Conference on Multimedia and Archiving Systems III, Nov. 1998.

[8] H. Wang and Shih-Fu Chang, "A Highly Efficient System for Automatic Face Region Detection in MPEG Video Sequences", IEEE Trans. on Circuits and Systems for Video Technology, special issue on Multimedia Systems and Technologies, Vol. 7, No. 4, Aug. 1997, pp. 615-628.

[9] H. Wang, H. S. Stone, and S.-F. Chang, "FaceTrack: Tracking and Summarizing Faces from Compressed Video", SPIE Multimedia Storage and Archiving Systems IV, 19-22 Sept, Boston".

[10] Y. Yacoob and M. J. Black, "Parameterized Modeling and Recognition of Activities", ICCV, 1998, pp120-127.

[11] J.K. Aggarwal and Q. Cai, "Human Motion Analysis: A Review," Computer Vision and Image Understanding, Vol.73, No.3, pp. 428-440, March 1999.

[12] D.M. Gavrila, "The Visual Analysis of Human Movement: A Survey", Computer Vision and Image Understanding, Vol.73, No.1, pp. 82-98, January 1999.

[13] M. Walter, S. Gong, A. Psarrou, "Learning Stochastic Temporal Models of Human Behaviour", Proc. IEEE International Workshop on Modelling People, Corfu,1999.

[14] K. Rangarajan, W. Allen, M. Shah, "Matching Motion Trajectories Using Scale-Space", Pattern Recognition, Vol 26, No 4, pp 595-610.

[15] R. Cutler and L. Davis, "Real-Time Periodic Motion Detection, Analysis and Applications", CVPR, 1999, pp. 326-332.

[16] C. Curio, J. Edelbrunner, T. Kalinke, C. Tzomakas, W. von Seelen, "Walking Pedestrian Recognition", ITSC, 1999, pp. 292-297.

[17] M. Kirby and L. Sirovich, "Application of the Karhumen-Loeve Procedure for the Characterization of Human Faces", IEEE PAMI, Vol 12(1), 1990, pp.103-108.

[18] M. Turk and A. Pentland, "Face Recognition Using Eigenfaces", CVPR 1991, pp. 586 -591.

ELEVIEW: An Active Elevator Video Surveillance System

Hui Shao, Liyuan Li, Ping Xiao and Maylor K.H. Leung
School of Computer Engineering, Nanyang Technological University
Singapore 639798
asmkleung@ntu.edu.sg

Abstract

In this paper, a novel study for an automated scene interpretation system, named ELEVIEW, is reported to outline the design of system. It is motivated by the reported crimes that happen inside elevators. The main goal is to investigate techniques that make an ordinary elevator monitoring system intelligent, i.e. see the scene and understand actions that are occurring. The system could filter out normal actions and trigger an alarm once abnormal events are detected. The paper focuses on the system overview, segmentation techniques as well as scenarios classification. A double thresholded segmentation is employed to enhance the segmentation outcomes. This paper mainly presents an overview of the system and significant results so far achieved.

1 Introduction

The goal of video surveillance systems for security applications is to identify people and their activities of interest in real-time in different kinds of environment. With the advent of relevant hardwares, VSAM (Video Surveillance and Monitoring) techniques [1, 2] have become affordable to assist security staff to monitor the military, commercial and other restricted area. Apart from the above usage, it can also be used in sport analysis, dance choreography, sign translation and gesture-driven user interface [3].

Since machine understanding of video has become attractive, VSAM will be playing a more and more important role in the future of security. Based on this view-point, the elevators, an arena of various abnormal activities, would be an ideal place for automatic monitoring. Nowadays, public area monitoring systems are all based on the *closed-circuit television (CCTV)* system that needs the supervision of security personnel. The system itself has no conscience on what is happening. It is just a sensor merely capturing and recording the scene. The pictures obtained from the CCTV are meaningless without the interpretation of a human being. Furthermore, security personnel have to stay in front of the screen all day, since something abnormal would take only a few minutes or even less to complete.

As an important component of motion analysis [4], human motion plays a key role in scenario recognition of a surveillance system. The study of human motion, specially walking in frontal parallel direction to the camera, has been extensively used as the basis of surveillance systems [1, 5, 6]. Pfinder [7] and W^4 [2] are the two of the typical real-time human body tracking and understanding systems. For the researches mentioned above, researchers [8, 9] have worked towards obtaining higher-level description, usually employing a linguistic approach for representing motion concepts [10]. In this study, scenarios used to define the occupant motions inside an elevator is depicted via a classification tree (see Figure 6).

The remainder of this paper is mainly organized to depict the work done to date: section 2 introduces the problem domain and presents a brief overview of system. Section 3 gives the application of a novel integrated method to segment the moving objects. Scenario classification is discussed in section 4, which is followed by a summary appeared in section 5.

2 System Overview
2.1 Problem Domain

The activities inside an elevator can be classified according to certain typical scenario types. A scenario is a sequence of expected activities (or sub-scenarios). For example, one would expect a murderous scenario to have the scene of at least two people in an elevator, with weapon swinging above the head, bloodstain on the wall and/or the victim falling down.

2.2 Scenario Classification

The more common and urgent actions inside an elevator can be generalized into forms of *still*, *fighting*,

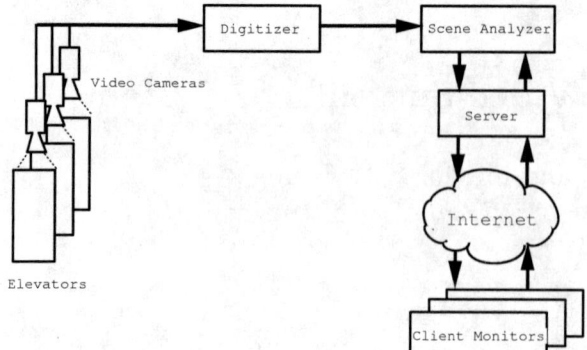

Figure 1: **System configuration.**

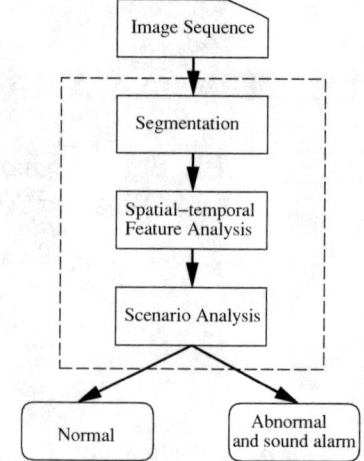

Figure 2: **Scene analysis process.**

vandalizing as well as *overstaying* as follows:

Still (normal): little or no motion is detected.

Fighting (suspicious): a lot of movements are detected.

Vandalizing (stain): a mess is left behind.

Overstaying: occupants overstay longer than an acceptable period.

2.3 System Configuration

The configuration of the monitoring system is proposed in Figure 1. ELEVIEW consists of five main parts, the video camera device, digitizer, scene analyzer, client monitor and Internet based web server. A video camera is used to capture the scene of an elevator. The digitizer transforms the analogy signal into digital data that can be processed by the scene analyzer, which is the core part of the system with intelligent processing to understand human activities. The analyzer would send an alarm to the security personnel when abnormal activities are detected. The client monitor can selectively request the server to send live image sequence or periodic update of any particular scene. The scene analysis process consists of three components (see Figure 2), namely, *Segmentation, Spatial-Temporal Feature Extraction* and *Scenario Analysis*. Images are digitized in resolution of 144×180 (row $\times$ column) pixels with 256 gray levels. Segmentation filters and separates moving regions/objects from static background. Spatial-temporal feature extraction extracts spatial and dynamic aspects of motion features, which are useful for human motion analysis inside the elevator. Scenario analysis analyzes activities inside the elevator through high-level description of human activities of scenarios.

3 Segmentation of Moving Objects

Segmentation is the first process to start off the analysis. Its task here is to obtain the moving regions/objects from current frames. The difference picture technique used commonly in computer vision is adopted in this work. Conventional moving objects segmentation is carried out by a direct pixel-by-pixel background subtraction. In this study, a novel segmentation approach is applied in order to obtain robust results.

3.1 Intensity Difference with Texture Information

Conventional direct intensity subtraction is found not robust enough to tackle the problems such as flickering light source, moving shadows of people, or merging of moving objects with background due to similar gray level values. Li and Leung [11] found that the texture structures of background and foreground would remain unchanged.

3.2 Texture Difference Measure

Since the edge information is less sensitive to light changes, the gradient is used to derive the local texture difference. Let $p = (x, y)$ be a point in an image plane, $B(p)$ the reference background, and $F(p)$ the current frame. Also let ∇B and ∇F represent the gradient vectors (B_x, B_y) and (F_x, F_y) respectively. Here the subscripts denote the partial derivative operators. At each point, the cross-correlation of gradient vectors of the background and current frame can be defined as

$$C_{bf} = \nabla B \cdot \nabla F = \|\nabla B\| \cdot \|\nabla F\| \cos\theta. \quad (1)$$

where θ is the angle between the vectors. Similarly the auto-correlations of gradient vectors of the background and current frame can be defined as

$$C_{bb} = \nabla B \cdot \nabla B = \|\nabla B\|^2 \qquad (2)$$

and

$$C_{ff} = \nabla F \cdot \nabla F = \|\nabla F\|^2 \qquad (3)$$

and we have

$$C_{bb} + C_{ff} \geq 2C_{bf} \qquad (4)$$

If the neighborhood of the point p is not covered by moving object in current frame, the local texture features of this region in the background and current frame would be similar, i.e. there are no great differences in length and direction of ∇B and ∇F. Hence, one has $C_{bb} + C_{ff} \approx 2C_{bf}$. On the other hand, if a background region is covered by a moving object in current frame, there would usually be large differences between the background and current frame textures since they are from regions of different objects. In this case, the gradient vectors ∇B and ∇F would be different in both length and direction. Then $2C_{bf}$ would become much smaller than $C_{bb} + C_{ff}$. Let's define a local measure of the neighborhood gradient difference as

$$R_t(p) = 1 - \frac{\sum_{u \in \mathcal{M}_p} 2C_{bf}(u)}{\sum_{u \in \mathcal{M}_p} (C_{bb}(u) + C_{ff}(u))} \qquad (5)$$

where $\mathcal{M}_p$ denotes the 5×5 neighborhood centered at p. Since $\|\theta\| > 90°$ would indicate much difference between the vectors, one can normalize $R_t(p)$ by setting $R_t(p) = 1$ (the maximum value) whenever $R_t(p) > 1$. This gradient difference is robust to light changes and noise. Since light changes would only cause the change of contrast or the lengths of the gradient vectors of background and current frame, the directions of the vectors remain intact. Noise usually creates perturbation in both the length and direction of the vectors, but the effect can be averaged out. Not unless exceptionally large contrast is created, $2C_{bf}$ would not be very much different from $C_{bb} + C_{ff}$ if the background region is not covered in the current frame. Hence $R_t(p)$ can be made use of to segment moving objects from background.

3.3 Integration of Texture/Intensity Difference

The texture and intensity differences can complement each other. They are two different views to the difference between the current frame and background. A better segmentation of the moving objects can therefore be achieved by integrating the information from these two sources. Two integration approaches for the intensity and texture difference were proposed as *Weighted Integration (WI)* and *Minimized Energy Integration (MEI)*, respectively. The details can be found in [11]. For both WI and MEI, the values noted as $D(p)$ have been normalized to be in the range from 0 and 1. Hence, the mid-point (0.5) is naturally selected to screen out the moving regions.

3.4 Segmentation Enhancement

The segmentation outcome discussed above stems from the scheme in which solely one threshold is utilized. The difference image is thresholded to keep only locations of significant changes. However, the experiment result was observed not satisfactory, i.e. the lower part of the human body is missing frequently due to the weak signals of intensity and/or texture difference (see Figure 3, (e)-(h), (m)-(p)). In order to segment the occupants better, the hysteresis thresholding is applied to strengthen the segmentation results. That means the difference picture is initially thresholded at a higher threshold. The threshold is later decreased to an acceptable level.

3.4.1 The Upper Threshold (T_d^{high})

Since the segmented moving regions are smaller than what are expected, a region expansion process is then needed. The mid-point (0.5) is regarded as the upper threshold T_d^{high} for the difference output. In short, if any pixel of the difference image has value above T_d^{high}, it is treated as a foreground pixel immediately. Let S_{high} be the set of such pixels.

3.4.2 The Lower Threshold (T_d^{low})

After thresholding with T_d^{high}, a set S_{low} can be collected as groups of connected pixels such that in each group, each pixel (p) would satisfy ($D(p) > T_d^{low}$) and ($D(p) \leq T_d^{high}$) and (at least one pixel from the group is an immediate neighbor of a pixel in S_{high}). The final thresholded image $\mathcal{P}$ is consequently defined as

$$\mathcal{P} = S_{high} \bigcup S_{low} \qquad (6)$$

From observations of the WI and MEI difference histograms (see Figure 4 and Figure 5), one can approximate the background noises as a Gaussian distribution pattern. Since the background should be removed, one can compute the value, $d_{bg}^{99\%}$, on the horizontal axis that corresponds to 99% area of the background Gaussian distribution as $\mu + 3 \times \sigma$. The mean value and

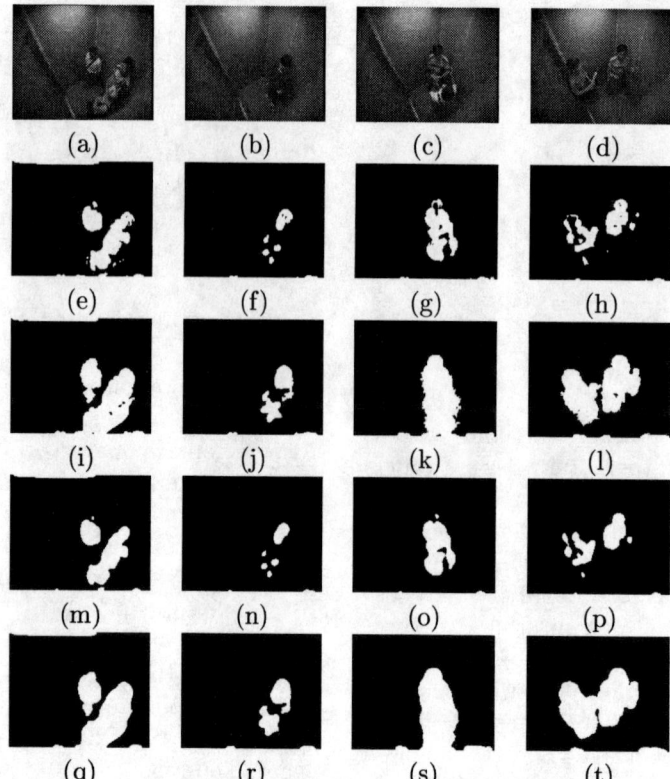

Figure 3: Comparison of segmentation results: (a) to (d): Seq1:4, Seq2:288, Seq3:100 and Seq4:10; (e) to (h): WI with T_d^{high}, (i) to (l): WI with T_d^{high} and T_d^{low}, (m) to (p): MEI with T_d^{high}, (q) to (t): MEI with T_d^{high} and T_d^{low}.

standard deviation can be estimated from the samples randomly selected from the image sequences. The estimation indicates that $\mu_{WI} = 7.93$ and $\sigma_{WI} = 5.17$. Consequently $d_{bg_{WI}}^{99\%} = \mu_{WI} + 3 \times \sigma_{WI} = 23.44$ or 0.09 when it is mapped back to the range of $[0,1]$. The computation for MEI produced $\mu_{MEI} = 7.27$ and $\sigma_{MEI} = 2.29$. The corresponding $d_{bg_{MEI}}^{99\%}$ is 14.14 or 0.06. Since the computation of $D(p)$ would take the average on p's neighborhood (i.e. affected very much by its neighbors), T_d^{low} has to be larger than $d_{bg_{WI}}^{99\%}$ or $d_{bg_{MEI}}^{99\%}$ but less than 0.5. In this work, it is set empirically as

$$T_d^{low} = 0.25 \qquad (7)$$

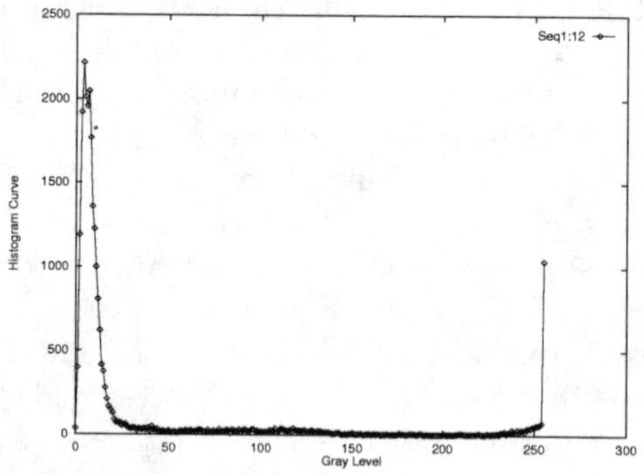

4 Scenario Analysis

As mentioned in Section 2.1, the scene analysis process consists of segmentation, spatial-temporal feature extraction and scenario analysis. In addition to the segmentation operation, other components will be introduced in the rest of this paper.

Figure 4: WI histogram curve of Seq1:12.

4.1 Spatial-temporal Feature Extraction

In this study, two types of features are investigated, namely, the spatial and temporal features. The spatial features capture the spatial shape information on the regions of interest whereas temporal features characterize the motion pattern.

70

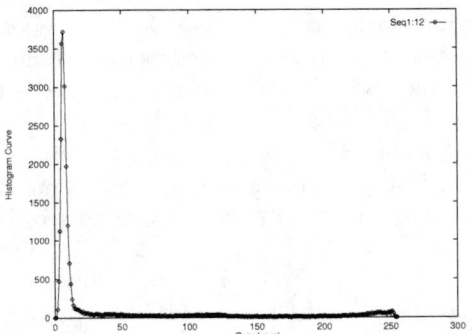

Figure 5: MEI histogram curve of Seq1:12.

4.1.1 Door Status Detection

The difference picture obtained from subtraction with the background is used for door detection. When the door is closed, the intensity median of the difference picture will be relatively stable and stay low since most regions are belong to the background. However, the intensity median would become much larger when the door is opened since more light would come into the elevator.

4.1.2 Occupancy Detection

Whether an elevator is empty or not can be achieved by measuring the size/area of a small child during the installation time.

4.1.3 Movement Aggressiveness

The temporal feature of movement aggressiveness (AG) is used to reveal abnormal activities caused by excessive movement. This can be computed by comparing the moving regions against the occupant area as AG_m, or measuring the changes of bounding box (i.e. changes of posture) as AG_b. Then AG will be taken as the maximum value of them, i.e. $AG = \max(AG_m, AG_b)$. The movement aggressiveness related to the moving regions (AG_m) is computed as

$$AG_m = \frac{\mathcal{M}_m^T}{\mathcal{M}_o^T} \qquad (8)$$

where $\mathcal{M}_o^T$ (see Figure 3) is the mask of occupants of the ith frame and $\mathcal{M}_m^T$ is a mask of the moving regions between the $(i-1)^{th}$ and i^{th} frames. AG_m measures the ratio of positional change of two consecutive frames. The value can range from zero to slightly larger than one, a large value of AG_m might signal an abnormal situation.

A bounding box is defined as the smallest rectangle enclosing a detected object. An object may contain one or more occupants due to merging. The shape of a bounding box can be described using its vertical height (H) and horizontal width (W). H and W change with the movement of the occupant. Consider the current frame (i) in a sequence. Let ΔH_{ik} be the height difference of the kth object bounding box with respect to the previous frame. Also let ΔW_{ik} and ΔA_{ik} be the width and area differences of the kth object bounding box with respect to the previous frame. AG_b is computed as follows:

$$\Delta H_{ik} = \frac{|H_{ik} - H_{(i-1)k}|}{H_{ik}} \qquad \Delta W_{ik} = \frac{|W_{ik} - W_{(i-1)k}|}{W_{ik}}$$
$$(9)$$

$$\Delta A_{ik} = \frac{|A_{ik} - A_{(i-1)k}|}{A_{ik}} \qquad (10)$$

The maximum differences are obtained through

$$\Delta H_{ik} = \max_k \{\Delta H_{ik}\} \qquad \Delta W_{ik} = \max_k \{\Delta W_{ik}\} \qquad (11)$$

$$\Delta A_{ik} = \max_k \{\Delta A_{ik}\} \qquad (12)$$

with $k = 1, 2, \ldots, K$ where K is the number of the matched objects found in the ith frame. Finally, the AG_{bi} for the ith frame is defined as

$$AG_{bi} = \max\{\Delta H_i, \Delta W_i, \Delta A_i\} \qquad (13)$$

4.2 Head Top Points Dectecion

The head top point is defined as the highest peak of an object detected in the segmented edge region picture. For example, the head top point can be tracked to determine whether an occupant has overstayed for an abnormal long period. A timer starts to tick when an occupant is to be tracked. It stops when the occupant goes out of the elevator. Currently a novel model-to-image *Line Hausdorff Distance (LHD)* is being applied to localize head top points based on the head boundary information.

4.3 Scenario Analysis

Scenario classification is a bi-directional classification process. It includes bottom-up feature collection and top-down verification processes. In this research, four kinds of scenarios are investigated, namely, suspicious scenario, overstaying scenario, stain scenario and normal scenario. These scenarios are described based on the spatial and temporal features discussed earlier. Figure 6 shows a classification tree of human activities inside an elevator. The elevator has two statuses: door closed and opened. Generally suspicious activities would only happen when the elevator door is closed so

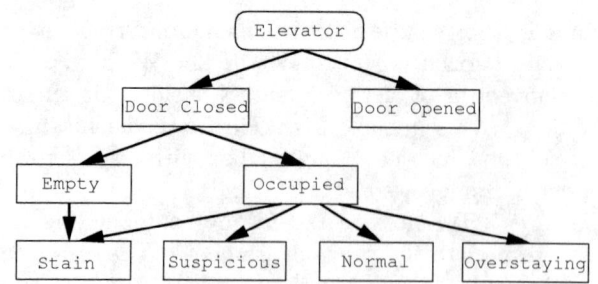

Figure 6: A classification tree of human activities inside an elevator.

as to avoid to be seen. This is the first step to start detection. When the door is detected closed, the system would determine whether the elevator is empty or not. The analysis will focus mainly on when the elevator is not empty. A revised tree is currently being investigated to add more intelligence to the system. The criteria for scenario classification are listed as:

Suspicious When excessive movement is sensed over long period, it is classified as suspicious. These can be detected by keeping track of the movement aggressiveness value AG.

Overstaying An occupant should not stay in an elevator for an abnormally long period. Otherwise it is classified as overstaying. This can be tracked by using the head top points. For example, If the overstay period is set to 20 frames, the bell will be rung at the 21st frame if the object has stayed for more than 20 frames.

Stain An object, which is found not touching the elevator bottom but hanging on wall, is considered suspicious since it could be bloodstain or stain caused by vandalism.

Normal The scene with no or little motion is classified as normal. In general, no abnormality occurs when the occupants keep still.

5 Summary

A study for exploring new techniques for an automated scene interpretation system, named ELEVIEW, is reported to outline the design of system. The proposed system aims to add intelligence to the ordinary *CCTV* system installed in elevators. In addition to the research on the scenarios inside the elevators, a novel segmentation method is applied here to enhance the initial stage of the system analysis. The segmentation does not use only intensity difference cue, but also uses the texture difference existing in between the background and foreground objects. The results show that the segmentation using WI and MEI is smoother and more robust than that using conventional background removal method. In order to strengthen the final results, a double measure thresholds scheme is utilized here to segment the moving objects better.

In our future work, we will be focusing on the study of feature extraction, i.e. the positioning of head-top and other human body parts. Based on the location of some critical human body parts, a tracking can be done in order to obtain the human body moving trajectory and therefore facilitate the scenario understanding.

Acknowlegments

The authors would like to thank CGIT from NTU for providing the experiment facilities.

References

[1] R. T. Collins and A . J. Lipton and T. Kanade, "A System for Video Surveillance and Monitoring," *Proceedings American Nuclear Society (ANS) 8th International Topical Meeting on Robotics and Remote Systems*, Pittsburgh, PA, 1999.

[2] I. Haritaoglu and D. Harwood and L. S. Davis, "W^4: Who? When? Where? What? A Real-time System for Detecting and Tracking People", *Proc. Int. Conf. Face and Gesture Recognition*, Nara, Japan, 1998.

[3] D. M. Gavrila and L. S. Davis, "3-D Model-based Tracking of Humans in Action: a Multiview Approach," *Proc. IEEE Computer Society Conf. CVPR'96*, pp. 73-80, 1996.

[4] J. K. Aggarwal and Q. Cai and W. Liao and B. Sabata, "Articulated and Elastic Non-rigid Motion: A Review", *Proc. Workshop on Motion of Non-Rigid And Articulated Objects*, pp. 2-14, 1994.

[5] M. K. Leung and Yee-Hong Yang, "First Sight: A Human Body Outline Labeling System", *IEEE Trans. Pattern Analysis and Machine Intelligence*, Vol. 17(4), pp. 359-377, 1995.

[6] L. W. Cambell and A. F. Bobick, "Recognition of Human Body Motion Using Phase Space Constraints", *Proc. 5th International Conference on Computer Vision*, pp. 624-630, 1995.

[7] C. Wren, A. Azarbayejani, T. Darrell and A. Pentland, "Pfinder: Real-time Tracking of the Human Body," *IEEE Trans. Pattern Analysis and Machine Intelligence*, Vol. 19(7), pp. 780-785, 1997.

[8] Y. Ivanov, C. Stauffer and A. Bobick, "Video Surveillance of Interactions" *2nd IEEE Workshop on Visual Surveillance*, pp 82-89, 1999.

[9] F. Brémond and M. Thonnat, "Analysis of Human Activities Described By Image Sequences", *Proc. 10th Intl. FLAIRS Conf.*, Florida, USA, 1997,

[10] P. Morasso and V. Tagliasco, *Human Movement Understanding: From Computational Geometry to Artificial Intelligence*, North-Holland, 1986,

[11] L. Li and M. K. H. Leung, "Robust Background Difference", *Proc. CVPRIP'2000*, 2000

Robust Head Motion Computation by Taking Advantage of Physical Properties

Zicheng Liu, Zhengyou Zhang
Microsoft Research, One Microsoft Way, Redmond, WA 98052, USA
{zliu,zhang}@microsoft.com

Abstract

Head motion determination is an important problem for many applications including face modeling and tracking. We present a new algorithm to compute the head motion between two views from the correspondences of five feature points (eye corners, mouth corners, and nose tip), and zero or more additional image point matches. The algorithm takes advantage of the physical properties of the feature points such as symmetry, and it significantly improves the robustness of head motion estimation. This is achieved by reducing the number of unknowns to estimate, thus increasing information redundancy. This idea can be easily extended to any number of feature point correspondences.

Keywords: robust motion estimation, head motion, face modeling, physical property, symmetry.

1 Introduction

In this paper, we present a new algorithm to compute the head motion between two views from the correspondences of five feature points including eye corners, mouth corners and nose tip, and zero or more additional image point matches. This is useful for applications such as face modeling and tracking systems where the feature points can be obtained. For example, the user can mark the feature points on the two images [5], or the user marks the feature points on one image while their correspondences are tracked on the other image [2], or the feature points are extracted and tracked automatically [3].

If the image locations of these feature points are precise, one could use five-point algorithm to compute camera motion. However, this is usually not the case in practice. A human in general cannot mark the feature points with high precision. A tracking algorithm may not result in perfect matches either. When there are errors, a five-point algorithm is not robust even when refined with a bundle adjustment technique. The key idea of our work is to use the physical properties of the feature points to improve the robustness. In this paper, we exploit the property that a face is almost

symmetric to reduce the number of unknowns. Additionally, we put reasonable lower and upper bounds on the nose height and represent the bounds as inequality constrains. As a result, the algorithm becomes significantly more robust.

Even though in this paper we only describe our algorithm in the case of five feature points, it is straightforward to extend the idea to any number (less than or bigger than 5) of feature points for robustness improvement. For example, we can drop the nose tip, or add outer eye corners.

In Section 2, we describe the motion estimation algorithm from five feature points only. In Section 3, we extend the algorithm to incorporate other image point matches obtained from image matching methods. The experiment results are shown in Section 4.

2 Head motion estimation from five feature points

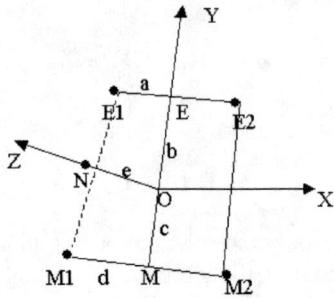

Figure 1. The new coordinate system Ω_0.

Although, as we said earlier, our approach can be extended to deal with different number of feature points by taking advantage of their physical properties, we show in this paper how to implement this by using five feature points on a face. The feature points are the left eye corner, right eye corner, left mouth corner, right mouth corner, and nose tip, which are denoted by E_1, E_2, M_1, M_2, and N, respectively (See Figure 1). Denote E as the midpoint of E_1E_2

and M the midpoint of $M_1 M_2$. Notice that human faces exhibit some strong structural properties. For example, left and right sides are very close to being symmetric about the nose; eye corners and mouth corners are almost coplanar. We therefore make the following reasonable assumptions: NM is perpendicular to $M_1 M_2$, NE is perpendicular to $E_1 E_2$, and $E_1 E_2$ is parallel to $M_1 M_2$.

Let π be the plane defined by E_1, E_2, M_1 and M_2. Let O denote the orthogonal projection of point N on plane π. Let Ω_0 denote the coordinate system which is originated at O with ON as the Z-axis, OE as the y-axis; the x-axis is defined according to the right hand system (See Figure 1). In this coordinate system, based on the assumptions mentioned earlier, we can define the coordinates of E_1, E_2, M_1, M_2, N as $(-a, b, 0)^T$, $(a, b, 0)^T$, $(-d, -c, 0)^T$, $(d, -c, 0)^T$, $(0, 0, e)^T$, respectively. Thus, we only need 5 parameters, $\{a, b, c, d, e\}$, to define these five points in this local coordinate system. This is to be compared with 9 parameters required for generic five points after choosing an appropriate local coordinate system. We have therefore reduced the number of unknowns by four.

Let $\mathbf{t}$ denote the coordinates of O in the camera coordinate system, and $\mathbf{R}$ the rotation matrix whose three columns are vectors of the three coordinate axis of Ω_0. For each point $\mathbf{p} \in \{E_1, E_2, M_1, M_2, N\}$, its coordinate under the camera coordinate system is $\mathbf{R}\mathbf{p} + \mathbf{t}$. We call $(\mathbf{R}, \mathbf{t})$ the *head pose transform*.

Given two images of the head under two different poses (assume the camera is static), let $(\mathbf{R}, \mathbf{t})$ and $(\mathbf{R}', \mathbf{t}')$ be their head pose transforms. For each point $\mathbf{p}_i \in \{E_1, E_2, M_1, M_2, N\}$, if we denote its image point in the first view by $\mathbf{m}_i$ and that in the second view by $\mathbf{m}'_i$, we have the following equations:

$$proj(\mathbf{R}\mathbf{p}_i + \mathbf{t}) = \mathbf{m}_i \qquad (1)$$

and

$$proj(\mathbf{R}'\mathbf{p}_i + \mathbf{t}') = \mathbf{m}'_i \qquad (2)$$

where $proj$ is the perspective projection. Notice that we can fix one of the a, b, c, d, e since the scale of the head size cannot be determined from the images. As is well known, each pose has 6 degrees of freedom. Therefore the total number of unknowns is 16, and the total number of equations is 20. If we instead use their 3D coordinates as unknowns as in any typical bundle adjustment algorithms, we would end up with 20 unknowns and have the same number of equations. By using the generic properties of the face structure, the system becomes over-constrained, making the pose determination more robust.

To make the system even more robust, we add an inequality constraint on e. The idea is to force e to be positive and not too large compared to a, b, c, d. This is obvious since

the nose is always out of plane π. In particular, we use the following inequality:

$$0 \le e \le 3a \qquad (3)$$

We chose 3 as the upper bound of e/a simply because it seems reasonable to us and it works well. The inequality constraint is finally converted to equality constraint by using penalty function.

$$P_{\text{nose}} = \begin{cases} e * e & \text{if } e < 0; \\ 0 & \text{if } 0 \le e \le 3a; \\ (e - 3a) * (e - 3a) & \text{if } e > 3a. \end{cases} \qquad (4)$$

In summary, based on equations (1), (2) and (4), we estimate a, b, c, d, e, $(\mathbf{R}, \mathbf{t})$ and $(\mathbf{R}', \mathbf{t}')$ by minimizing

$$\mathcal{F}_{\text{5pts}} = \sum_{i=1}^{5} w_i (\|\mathbf{m}_i - proj(\mathbf{R}\mathbf{p}_i + \mathbf{t})\|^2 \\ + \|\mathbf{m}'_i - proj(\mathbf{R}'\mathbf{p}_i + \mathbf{t}')\|^2) + w_n P_{\text{nose}} \qquad (5)$$

where w_i's and w_n are the weighting factors, reflecting the contribution of each term. In our case, $w_i = 1$ except for the nose term which has a weight of 0.5 because it is usually more difficult to locate the nose tip than other feature points. The weight for penalty, w_n, is set to 10. The objective function (5) is minimized using a Levenberg-Marquardt method [4]. More precisely, as mentioned earlier, we set a to a constant during minimization since the global head size cannot be determined from images. The initial values of $\{a, b, c, d, e\}$ are computed from an average face. The initial rotation matrices $\mathbf{R} = \mathbf{R}' = \mathbf{I}$, and the initial translation vectors $\mathbf{t} = \mathbf{t}' = [0, 0, 2]^T$; that is, we assume the head is in front of the camera and roughly facing toward it.

3 Incorporating image point matches

If we estimate camera motion using only the five user-marked points, the result is sometimes not very accurate because the markers contain human errors. In this section, we describe how to incorporate the image point matches (obtained by any feature matching algorithm such as the one described in [8]) to improve accuracy.

Let $(\mathbf{m}_j, \mathbf{m}'_j)$ $(j = 1, \dots, K)$ be the K point matches, each corresponding to the projections of a 3D point $\mathbf{p}_j$ according to the perspective projection (1) and (2). Obviously, we have to estimate $\mathbf{p}_j$'s which are unknown. Assuming that each image point is extracted with the same accuracy, we can estimate a, b, c, d, e, $(\mathbf{R}, \mathbf{t})$, $(\mathbf{R}', \mathbf{t}')$ and $\{\mathbf{p}_j\}$ $(j = 1, \dots, K)$ by minimizing

$$\mathcal{F} = \mathcal{F}_{\text{5pts}} + w_p \sum_{j=1}^{K} (\|\mathbf{m}_j - proj(\mathbf{R}\mathbf{p}_j + \mathbf{t})\|^2 \\ + \|\mathbf{m}'_j - proj(\mathbf{R}'\mathbf{p}_j + \mathbf{t}')\|^2) \qquad (6)$$

where $\mathcal{F}_{5\text{pts}}$ is given by (5), and w_p is the weighting factor. We set $w_p = 1$ by assuming that the extracted points have the same accuracy as those of eye corners and mouth corners. The minimization can again be performed using a Levenberg-Marquardt method.

This is quite a large minimization problem since we need to estimate $16 + 3K$ unknowns, and therefore it is computationally quite expensive especially for large K. Fortunately, as shown in [7], we can eliminate the 3D points using a first order approximation. The following term

$$\|\mathbf{m}_j - proj(\mathbf{R}\mathbf{p}_j + \mathbf{t})\|^2 + \|\mathbf{m}'_j - proj(\mathbf{R}'\mathbf{p}_j + \mathbf{t}')\|^2$$

can be shown to be equal, under the first order approximation, to

$$\frac{(\tilde{\mathbf{m}}'^T_j \mathbf{E} \tilde{\mathbf{m}}_j)^2}{\tilde{\mathbf{m}}^T_j \mathbf{E}^T \mathbf{Z}\mathbf{Z}^T \mathbf{E} \tilde{\mathbf{m}}_j + \tilde{\mathbf{m}}'^T_j \mathbf{E}\mathbf{Z}\mathbf{Z}^T \mathbf{E}^T \tilde{\mathbf{m}}'_j}$$

where $\tilde{\mathbf{m}}_j = [\mathbf{m}^T_j, 1]^T$, $\tilde{\mathbf{m}}'_j = [\mathbf{m}'^T_j, 1]^T$, $\mathbf{Z} = \begin{bmatrix} 1 & 0 \\ 0 & 1 \\ 0 & 0 \end{bmatrix}$, and $\mathbf{E}$ is the essential matrix to be defined below.

Let $(\mathbf{R}_r, \mathbf{t}_r)$ be the relative motion between two views. It is easy to see that

$$\mathbf{R}_r = \mathbf{R}'\mathbf{R}^T$$
$$\mathbf{t}_r = \mathbf{t}' - \mathbf{R}'\mathbf{R}^T\mathbf{t}$$

Furthermore, let's define a 3×3 antisymmetric matrix $[\mathbf{t}_r]_\times$ such that $[\mathbf{t}_r]_\times \mathbf{x} = \mathbf{t}_r \times \mathbf{x}$ for any 3D vector $\mathbf{x}$. The essential matrix is then given by

$$\mathbf{E} = [\mathbf{t}_r]_\times \mathbf{R}_r \qquad (7)$$

which describes the epipolar geometry between two views [1].

In summary, the objective function (6) becomes

$$\mathcal{F} = \qquad \mathcal{F}_{5\text{pts}}$$
$$+ w_p \sum_{j=1}^{K} \frac{(\tilde{\mathbf{m}}'^T_j \mathbf{E} \tilde{\mathbf{m}}_j)^2}{\tilde{\mathbf{m}}^T_j \mathbf{E}^T \mathbf{Z}\mathbf{Z}^T \mathbf{E} \tilde{\mathbf{m}}_j + \tilde{\mathbf{m}}'^T_j \mathbf{E}\mathbf{Z}\mathbf{Z}^T \mathbf{E}^T \tilde{\mathbf{m}}'_j} \qquad (8)$$

Notice that this is a much smaller minimization problem. We only need to estimate 16 parameters, the same as in the five-point problem (5), instead of $16 + 3K$ unknowns.

To obtain a good initial estimate, we first use only the five feature points to estimate the head motion by using the algorithm described in Section 2. Thus we have the following two step algorithm:

Step1. Set $w_p = 0$. Solve minimization problem 8.

Step2. Set $w_p = 1$. Use the results of step1 as the initial estimates. Solve minimization problem 8.

Notice that we can apply this idea to the more general cases where the number of feature points is not five. For example, if there are only two eye corners and mouth corners, we'll end up with 14 unknowns and $16 + 3K$ equations. Other symmetric feature points (such as the outside eye corners, nostrils, etc) can be added into equation 8 in a similar way by using local coordinate system Ω_0.

4 Results

In this section, we show some test results to compare our new algorithm with the traditional algorithms. Since there are multiple traditional algorithms, we chose to implement the algorithm as described in [6]. It works by first computing an initial estimate of the head motion from the essential matrix [1], and then re-estimate the motion with a nonlinear least-squares technique.

We have run both the traditional algorithm and our new algorithm on many real examples. We found many cases where the traditional algorithm fails while our new algorithm successfully results in reasonable camera motions. Figure 2 is such an example. The top row shows a pair of images with five markers each. The middle row shows the image matching points which are obtained by using a feature-based image matching algorithm [8]. The green lines are the motion vectors of the matches. The motion computed by the traditional algorithm is completely bogus, and the 3D reconstructions give meaningless results. But our new algorithm gives a reasonable result. We generate 3D reconstructions based on the estimated motion, and perform Delauney triangulation. The left image at the bottom row shows the texture mapped triangles at a new pose (the top left image is used as the texture map), and the image on the right shows its wire frame.

In order to know the ground truth, we have also performed experiments on artificially generated data. We arbitrarily select 80 vertices from a 3D face model (Figure 4) and project these vertices on two views (the head motion is eight degrees apart). The image size is 640 by 480 pixels. We also project the five 3D feature points (eye corners, nose tip, and mouth corners) to generate the image coordinates of the markers. We then add random noises to the coordinates (u, v) of both the image points and the markers. The noises are generated by a pseudo-random generator subject to Gaussian distribution with zero mean and standard deviation ranging from 0.4 to 1.2 pixels. Notice that we add noise to the marker's coordinates as well. The results are plotted in Figure 3. The blue curve shows the results of the traditional algorithm and the red curve shows the results of our new algorithm. The horizontal axis is the standard deviation of the noise distribution. The vertical axis is the difference between the estimated motion and the actual motion. Both estimated and actual translation vectors are normalized, and the difference is measured as their Euclidean distance. The difference between two rotations is measured as the Euclidean distance

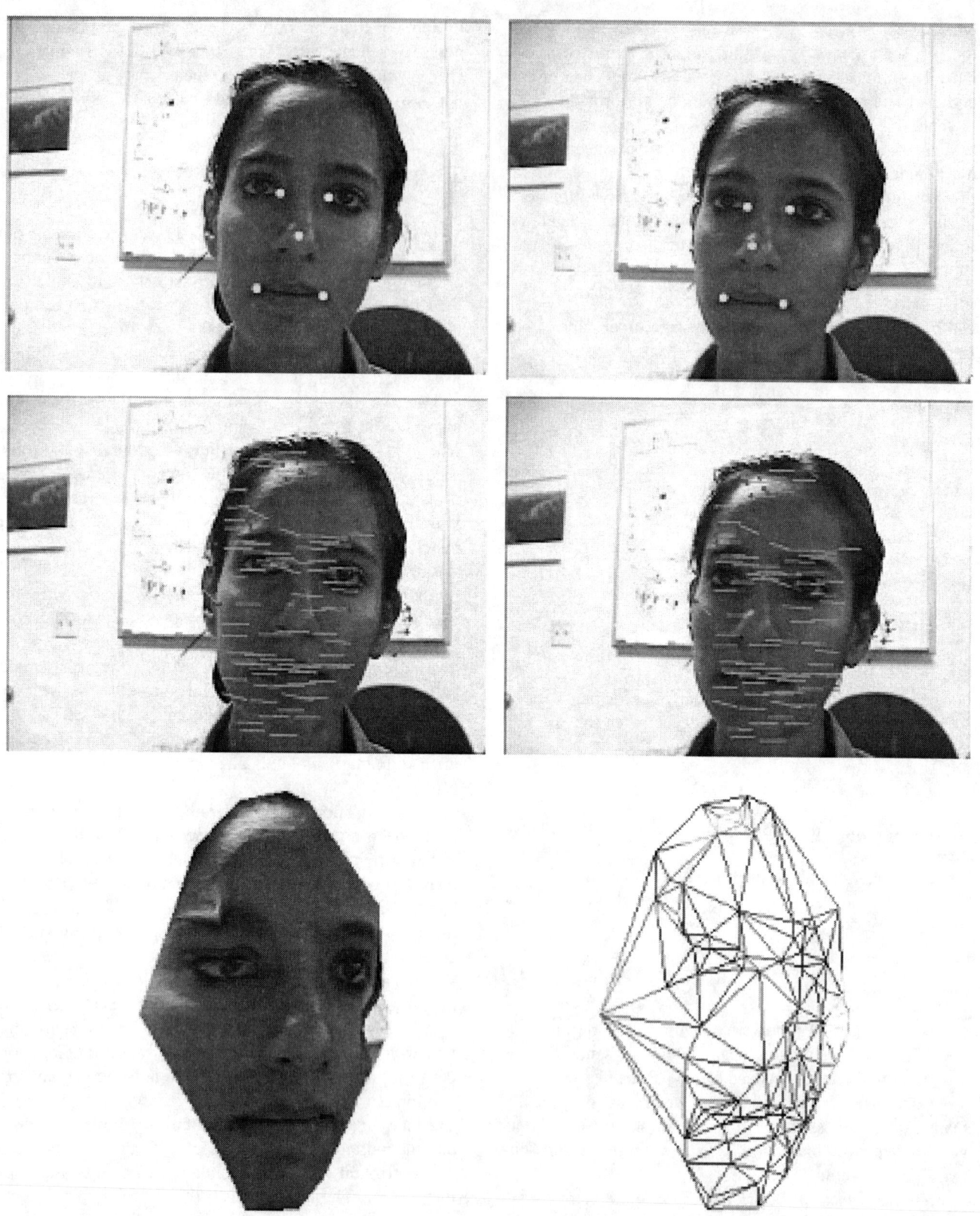

Figure 2. Top row: a pair of images with five markers. Middle row: image matching points. Bottom row: a novel view of the 3D reconstruction of the image matching points with the head motion computed by our new algorithm.

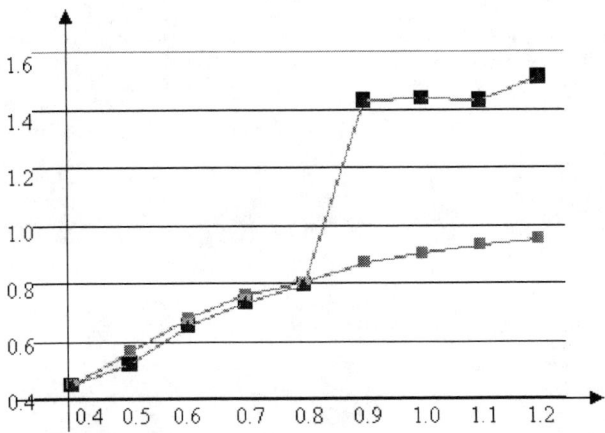

Figure 3. Comparison of the new algorithm with the traditional algorithm. The blue curve shows the results of the traditional algorithm and the red curve shows the results of our new algorithm. The horizontal axis is the standard deviation of the added noise (in pixels). The vertical axis is the error of computed head motion.

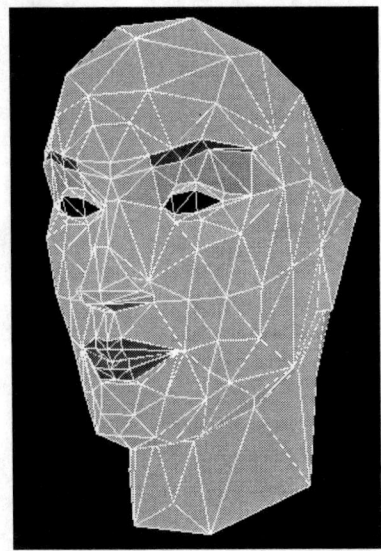

Figure 4. The face model used for experiments.

between the two rotational matrices. Figure 3 shows the average of combined motion errors from 20 random trials for each noise level. We can see that as the noise increases, the error of the traditional algorithm has a sudden jump at certain point, indicating algorithmic instability with respect to large noise. But the error with our new algorithm grows much more slowly.

5 Conclusion

We have developed a new head motion estimation algorithm which takes advantage of the physical properties, in particular the symmetry, of five human face features. The algorithm significantly improves the robustness over the traditional method. This is achieved by reducing the number of unknowns to estimate, thus increasing information redundancy. It can be applied to human face modeling and head tracking systems where the markers can be obtained either through user intervention or by using automatic feature detection algorithms. This algorithm can be easily extended to general cases where the number of feature points is not necessarily five. Furthermore, we have shown how to make full use of any additional image point matches, which do not have any particular semantic meaning, to further improve the accuracy.

References

[1] O. Faugeras. *Three-Dimensional Computer Vision: a Geometric Viewpoint*. MIT Press, 1993.

[2] P. Fua. Using model-driven bundle-adjustment to model heads from raw video sequences. In *International Conference on Computer Vision*, pages 46–53, Sept. 1999.

[3] T. S. Jebara and A. Pentland. Parameterized structure from motion for 3d adaptive feedback tracking of faces. In *Proc. CVPR*, pages 144–150, 1997.

[4] J. More. The levenberg-marquardt algorithm, implementation and theory. In G. A. Watson, editor, *Numerical Analysis*, Lecture Notes in Mathematics 630. Springer-Verlag, 1977.

[5] F. Pighin, J. Hecker, D. Lischinski, R. Szeliski, and D. H. Salesin. Synthesizing realistic facial expressions from photographs. In *Computer Graphics, Annual Conference Series*, pages 75–84. Siggraph, July 1998.

[6] Z. Zhang. Motion and structure from two perspective views: From essential parameters to euclidean motion via fundamental matrix. *Journal of the Optical Society of America A*, 14(11):2938–2950, 1997.

[7] Z. Zhang. On the optimization criteria used in two-view motion analysis. *IEEE Transactions on Pattern Analysis and Machine Intelligence*, 20(7):717–729, 1998.

[8] Z. Zhang, R. Deriche, O. Faugeras, and Q.-T. Luong. A robust technique for matching two uncalibrated images through the recovery of the unknown epipolar geometry. *Artificial Intelligence Journal*, 78:87–119, Oct. 1995.

Multiple Persons or Multiple Cameras I

Ray Carving with Gradients and Motion

Bradley Stuart and Yiannis Aloimonos
Computer Vision Laboratory
Center for Automation Research
Department of Computer Science and
Institute for Advanced Computer Studies
University of Maryland
College Park, MD 20742
{brad,yiannis}@cfar.umd.edu

Abstract

Recent developments in camera and computer technology have made multiple-camera systems less expensive and more usable. Using such systems, we can generate 3-D models of human activity for use in survelliance, as avatars, or for 3-D effects generation. Some approaches to model generation are voxel coloring, space carving, sihlouette intersection, and the combination of multiple stereo reconstructions.

Our attempt to overcome various shortcomings of the above approaches has led to the use of image derivatives and motion to determine the shape and motion of the activity in view. Direct computations of the gradient directions and the image motion normal to the gradient provide the information to generate a 3D + motion model consistent with all the image data. Data structures encode visibility information from each of the cameras surrounding the scene, allowing efficient determination of the subsets of measurements to be combined in a modified space-carving system.

The main conributions of this paper are the following: the development of a system for combining multiple image gradient measurements to determine the 3-D iso-brightness direction and its consistency, a system for combining multiple normal flow measurements to determine the motion normal to the iso-brightness direction, and a data structure based on the rays passing through the centers of projection and the image pixels, forming an unbounded projective grid through the space of the scene and allowing efficient determination and updating of scene point visibility.

Reconstructions of human motion using twenty cameras are presented.

1 Previous work

The question of how to generate new views of a three-dimensional (3-D) scene or object has been addressed in a number of different ways. The methods used fall into three rough categories, depending on whether the method uses only two-dimensional (2-D) data, builds a 2-D depth map ($2\frac{1}{2}$-D sketch), or generates new views from a full 3-D model of the scene.

1.1 Voxel carving

Much recent work has concerned the integration of a large number of views into a single 3-D object model. In the technique called space carving or voxel coloring, space is broken up into small cells (volume elements, or voxels) and the images of these cells are checked for consistency. Cells with inconsistent images are removed from the set of cells which constitute the object [2], [7], [10].

The issue of cell visibility can be addressed in a variety of ways. In [10], the camera configuration is controlled so that voxels are visited in near-to-far order. The space carving technique [7] allows arbitrary camera placement, but cameras are not used until they are passed by the plane sweeping through the scene. Generalized voxel coloring [2] maintains a "layered depth image" to store visibility information, and efficiently determine which voxels are revealed by the carving operation.

Different criteria for voxel consistency have also been used. The technique of volume intersection (shape from silhouettes) uses the logical AND of bit mask images to define consistent voxels. Seitz develops a color consistency test which is based on hue.

81

1.2 Morphing

Other approaches to the problem of generating new views from a given set of views are image interpolation and morphing. In image interpolation, interposed frames are created by smoothly varying pixel values between two or more source frames. Interpolation gives unsatisfactory results unless the source frames are very close together.

Morphing improves on this technique by imposing a mesh on the source frames, and moving control points on this mesh smoothly between the frames. Intermediate frames are generated by texture mapping each mesh cell as it moves, and varying the cell texture maps smoothly between the source frames. Defining the mesh requires the establishment of correspondences between points in the source images. Both image interpolation and morphing are strictly 2-D techniques; they do not perform well in the presence of occlusions and 3-D structure in the scene.

1.3 Stereo techniques

Stereo techniques use two or more images to compute 3-D structure in the scene. The epipolar constraint is used along with some technique (such as correlation) to find corresponding points on epipolar lines. Additional constraints are added to speed searching or deal with ambiguities. Examples of these are ordering constraints, inter-line constraints and depth smoothness constraints [13]. Once correspondences has been determined, the distance to a point in the scene is then an inverse function of the disparity in the image coordinates. Stereo techniques work well only with a limited range of camera separations. If the cameras are too close, images are too similar, disparities are small, and accuracy suffers. If the cameras are too widely separated, correspondences become difficult to find and the additional constraints employed to find them start to be violated more and more. Narayanan et al. [8] demonstrate techniques combining multiple stereo pairs and filling in the holes that appear behind a single depth map. In [12], optical flow values are back-projected onto these models and 3-D flow values are inferred. Alternately, the optical flow values can be used to augment incomplete 3-D models to make them more accurate.

1.4 Structure from motion

Structure-from-motion techniques use the relative motion between camera and scene to determine the depth of points in the scene [11]. The technique is based on motion parallax; when the camera translates relative to the scene, points near the camera have a greater apparent motion than points far away. This translational flow is compounded with the apparent motion due to rotation of the camera. This rotation gives no depth information, but its confusion with the translation makes the structure-from-motion problem considerably more difficult. Structure-from-motion techniques can be based on either optical flow or normal flow. Either way, the technique assumes that flows are due only to the motion of the camera; independent motion in the scene is not allowed. Furthermore, inaccuracies in the determination of camera motion lead directly to inaccuracies in the resulting depth map.

2 Camera setup and image formation

Up to sixteen cameras are installed on each of four walls, for a total of sixty-four cameras. Sixteen of these cameras are single-ccd RGB filtered color cameras; the rest are standard gray-scale cameras. All cameras send 8 bits per pixel of digital data at 640x480 pixels and 60 frames per second. Frame rates up to 85 fps are possible with a restricted region of interest. Data is fed from each camera to a dedicated video capture card; four such cards are installed in each of 16 Pentium II PC's. Each PC has one gigabyte of RAM to allow real-time capture of 3000 uncompressed frames per computer [3]. Projection matrices are computed by nonlinear optimization on up to 25 points of a large calibration object, or more accurately by the method detailed in [1].

3 Ray carving

The process of voxel carving typically begins with the division of space into cubes of some fixed size. This introduces restrictions on the shape and size of the space that can be carved, as well as restrictions on the resolution of the cameras that can be usefully employed.

3.1 Voxels or rays

Instead of splitting the scene arbitrarily with a regular grid, we base our division of space on the rays which pass through the pixels in each camera. Each camera defines a pyramid-shaped bundle of rays passing through the scene. Saito and Kanade [9] define a grid based on the rays from two selected cameras. Here, we do not define any special cameras, using the rays from all cameras equally. Furthermore, we use a continuous (floating point) rather than discrete representation of the intervals on the ray which are solid or transparent. Each filled interval is represented by a pair of numbers, the distances to the endpoints of the interval. These distances are adjusted as portions of the scene are carved away.

The carving method of Seitz [10] restricts the camera positions to be "occlusion compatible." This means that it must be possible to separate the scene from the cameras by a flat or concave surface, and sweep the surface out, processing voxels from near to far. Cameras which violate this ordering can be excluded until they are passed by the sweep surface. One of any pair of cameras which can "see" each other must be excluded until the sweep surface passes the line joining their centers. The asymmetries induced by excluding one camera can be alleviated by sweeping in multiple directions.

One chief advantage of the ray representation is that it provides a compact description of visibility. A point on a ray is visible if its distance is less than or equal to the distance to the beginning of the ray's first filled interval. Rays can be processed in arbitrary or random order; it is not necessary to sweep a surface through the scene.

3.2 Computing distances along rays

Carving efficiently using ray information requires the ability to convert image positions along the ray into 3-D distances and positions. All projective transformations of a line can be defined by specifying the transformations of only three points on that line. Using this fact, we can specify the transformations back and forth among 3-D coordinates along a ray, 2-D image coordinates in any other camera's view of that ray, and 1-D line coordinates of distance along the ray.

The visibility information for the ray is supplemented with the information needed to render the ray into all of the other views. This is analogous to the rendering of epipolar lines needed to determine stereo correspondences. The projective mapping between the ray in 3-D and its images in all the other views is completely determined by defining the mapping for three distinct points on the ray. The camera center is one such point, selected because it is shared by all rays in one image. For a 3×4 projection matrix $\tilde{\mathbf{P}} = [\mathbf{P}\tilde{\mathbf{p}}]$, where $\mathbf{P}$ is a 3×3 matrix and $\tilde{\mathbf{p}}$ is a 3×1 vector, Faugeras [4] gives its position as

$$\mathbf{C} = \mathbf{P}^{-1}\tilde{\mathbf{p}}. \qquad (1)$$

The second point selected is the vanishing point for the given ray — the vector parallel to the ray. Again from Faugeras, for homogeneous pixel coordinate $\mathbf{x}$ this direction is the homogeneous vector $[\mathbf{D}^{\mathsf{T}}0]^{\mathsf{T}}$, where

$$\mathbf{D} = \mathbf{P}^{-1}\mathbf{x}. \qquad (2)$$

The third point $\mathbf{U}$ is taken at a unit distance from the camera center, in the direction of the vanishing point. In homogeneous coordinates,

$$\mathbf{U} = \begin{bmatrix} \mathbf{C} \\ 1 \end{bmatrix} + \begin{bmatrix} \mathbf{D} \\ 0 \end{bmatrix}, \qquad (3)$$

where $\mathbf{D}$ has been scaled so that $\|\mathbf{D}\|_2 = 1$.

By projecting these three points into each of the other images, we are able to convert positions along the image of the ray directly into 3-D distances along the ray itself. Specifically, we define projectivities between the homogeneous image coordinates in the ith camera $\mathbf{c} = \tilde{\mathbf{P}}_i\mathbf{C}$, $\mathbf{d} = \tilde{\mathbf{P}}_i\mathbf{D}$, and $\mathbf{u} = \tilde{\mathbf{P}}_i\mathbf{U}$, and the homogeneous line coordinates $[0, 1]^{\mathsf{T}}$, $[1, 0]^{\mathsf{T}}$ and $[1, 1]^{\mathsf{T}}$, respectively. This projectivity will also map a point $\mathbf{x} = \tilde{\mathbf{P}}_i\mathbf{X}$ lying on the line to the line coordinate $[x_1, x_2]$, where $\frac{x_1}{x_2}$ is the distance from the camera center $\mathbf{C}$ defining this ray to $\mathbf{X}$. This projectivity is defined by the matrix product

$$\mathbf{L} = \begin{bmatrix} s_1 & 0 \\ 0 & s_2 \end{bmatrix} \begin{bmatrix} \mathbf{c} \times \mathbf{n} \\ \mathbf{n} \times \mathbf{d} \end{bmatrix}. \qquad (4)$$

The second matrix in this definition ensures that the points $\mathbf{c}$ and $\mathbf{d}$ map correctly, while the first defines the scale so that $\mathbf{u}$ maps to the point $[1, 1]$. The vector $\mathbf{n}$ in the second matrix is an arbitrary vector defining the null-space of the matrix L. For points $\mathbf{x} = \tilde{\mathbf{P}}_i\mathbf{X}$ on the ray, $\mathbf{Lx}$ gives the homogeneous coordinate measuring the 3-D distance from the camera center $\mathbf{C}$ to the point $\mathbf{X}$.

Requiring that the point $\mathbf{u}$ maps to the line coordinate $[1, 1]$ gives us the equation

$$\begin{bmatrix} s_1 & 0 \\ 0 & s_2 \end{bmatrix} \begin{bmatrix} \mathbf{c} \times \mathbf{n} \\ \mathbf{n} \times \mathbf{d} \end{bmatrix} \mathbf{u} \equiv \begin{bmatrix} 1 \\ 1 \end{bmatrix}$$

or

$$s_1[\mathbf{cnu}] = s_2[\mathbf{ndu}],$$

which is satisfied by setting $s_1 = [\mathbf{ndu}]$ and $s_2 = [\mathbf{cnu}]$.

Applying this projectivity to points which do not lie strictly on the line introduces errors when taking the resulting line coordinates as the depth values at the given point. The null-space of $\mathbf{L}$ defines the sets of points in the plane which project to the same line coordinates. For points not on the ray, we take their coordinate to be the same as the coordinate of the perpendicular projection onto the ray. To achieve this, we define the null-space vector to be the point at infinity in the direction perpendicular to the line containing $\mathbf{c}$ and $\mathbf{d}$. This is given by taking the cross product $\mathbf{c} \times \mathbf{d}$, and setting its third coordinate to zero. Thus

$$\mathbf{n} = [c_2d_3 - c_3d_2, -c_1d_3 + c_3d_1, 0]^{\mathsf{T}}.$$

Once the distance coordinate along the line is known, it is a simple matter to convert this to a 3-D coordinate. For any distance coordinate λ,

$$\mathbf{X}_\lambda = \begin{bmatrix} \mathbf{C} \\ 1 \end{bmatrix} + \lambda \begin{bmatrix} \mathbf{D} \\ 0 \end{bmatrix}, \qquad (5)$$

where again, $\mathbf{D}$ has been scaled to unity.

3.3 Carving along a ray

Ray carving requires projecting a given ray into all the other views. It is more computationally efficient to continue carving down the length of the ray until the visible end of the ray is no longer inconsistent. While updating the visible end of the current ray, moving its front point away from the camera, rays from the other cameras which view this front point are also updated. The carving algorithm proceeds as follows:

1. Determine the 3-D point x which is visible on this ray R as in (5).

2. Project the principal points $(\mathbf{C}, \mathbf{D}, \mathbf{U})$ of the ray R into the other views i, giving image coordinates $(r, c)_i$ and defining the matrices $\mathbf{L}(R)_i$ as in (4). Recall that $\mathbf{C}$ is common to all rays from a particular view; its projected coordinate (the epipole) can be saved. Saving other values for the images of the rays is memory intensive.

3. Image coordinates $(r, c)_i$ are used to access predefined rays S_i, the rays which view x in the other cameras.

4. Determine the depth boundaries at which the ray R crosses the edges of the pixels $(r, c)_i$ in each of the views. One of the views, i_{lub} defines the least upper bound on this depth.

5. If the pixels $(r, c)_i$ are inconsistent with the pixel defining the ray R (see below), the portion of the current ray between the visible point and the least upper bound defined above must be removed. The removed region extends onto the second least (unique) depth value found in the previous step.

6. In addition, the ray $S_{i_{lub}}$ defined in the view i_{lub} is also carved, removing from it the region defined by $S_{i_{lub}}$'s intersection with the pixel defining ray R.

7. The position x on ray R is updated, along with the definition of $S_{i_{lub}}$, the ray containing the portion which was carved away. Other rays S_i do not need to be recomputed, or carved, as the point x has moved to a new pixel only in view i_{lub}.

Since visibility information is encoded and updated for each ray, they can be visited in arbitrary order. Our choice is to visit all the rays defined by one image before moving on to the next. Cycling through the images continues until the scene converges and new points are no longer carved away.

4 Intensity gradients and edges in three dimensions

Typically, voxel carving algorithms have used color consistency as a test function in determining what portions of the scene need to be removed from the object model. Color has several advantages that a simple gray level does not. Color is less sensitive to variation in viewpoint, camera intensity response, and differences in lighting.

Gray-scale images are not without their own properties that can perform equally well. Intensity gradients can perform better than simple intensities, as they are less sensitive to the camera intensity response and differences in lighting. However, intensity gradients are not independent of viewpoint — a simple rotation of the camera in place will induce an opposite rotation of the gradient vectors. Furthermore, gradients which appear at a 3-D point in one image may not be present at all in another image, as at the occluding boundary of a smooth object. As we will see, these viewpoint dependent characteristics can be overcome by computing the 3-D iso-intensity contour.

Intensity gradients arise in images for several reasons. A surface in the scene may reflect or emit different intensities due to texture intrinsic to the surface or due to lighting differences such as shadows. An occluding boundary (discontinuity) in the scene produces a gradient when the occluding and occluded objects are of different colors. Specular reflections and other departures from a Lambertian reflectance model also produce intensity gradients. We do not deal with specularities here, but they could be addressed by including a lighting model of the scene, for example.

The gradient vector $dI/d\mathbf{x} = [I_x, I_y]^\top$ is the intensity change per unit length (usually a single pixel width) in the direction of steepest ascent. The normal to this vector is the direction in which the image intensity remains constant; the iso-intensity contour. Projecting the image intensity values back along their rays into three dimensions, we have a space-filling gradient field, without regard to where the physical surfaces lie. In Euclidean 3-D coordinates $\mathbf{X}$, this field has the brightness gradient

$$\frac{dB}{d\mathbf{X}} = k \begin{bmatrix} \frac{\partial x_1}{\partial X_1} & \frac{\partial x_2}{\partial X_1} \\ \frac{\partial x_1}{\partial X_2} & \frac{\partial x_2}{\partial X_2} \\ \frac{\partial x_1}{\partial X_3} & \frac{\partial x_2}{\partial X_3} \end{bmatrix} \frac{dI}{d\mathbf{x}}, \qquad (6)$$

where $k = dB/dI$, i.e. we assume a linear relationship between image intensity and scene brightness.

Using the 3×4 projection matrix $\mathbf{P}$ and homogeneous coordinates

$$\hat{\mathbf{X}}^\top = [\ \mathbf{X}^\top \ \ 1\ ],$$

we define

$$\hat{\mathbf{x}} = \mathbf{P}\hat{\mathbf{X}}.$$

84

The image of $\mathbf{X}$ is then at the 2-D coordinates

$$x_1 = \frac{\hat{x}_1}{\hat{x}_3}, x_2 = \frac{\hat{x}_2}{\hat{x}_3}. \qquad (7)$$

The derivatives of (6) are then given by

$$
\begin{aligned}
\frac{\partial x_i}{\partial X_j} &= \frac{\partial}{\partial X_j}\left(\frac{\hat{x}_i}{\hat{x}_3}\right) \\
&= \left(\frac{\partial \hat{x}_i}{\partial X_j}\hat{x}_3 - \frac{\partial \hat{x}_3}{\partial X_j}\hat{x}_i\right)/\hat{x}_3^2 \\
&= (P_{i,j}\hat{x}_3 - P_{3,j}\hat{x}_i)/\hat{x}_3^2.
\end{aligned}
$$

As in the 2-D case, the gradient vector is normal to the manifold of constant brightness. This is the iso-brightness plane, which must contain the direction of constant brightness lying on the actual surface in the scene.

Each of the views gives a measurement of the image gradient, and therefore a plane on which the iso-brightness direction must lie. With two such planes, a unique line is defined for the edge direction in 3-D. With more than two planes, we can use a fitting technique to find the best direction for the 3-D edge. Furthermore, we can use the quality of this fit to determine whether these image gradients do in fact define a consistent edge in 3-D.

The fitting technique used is a principal component analysis. For N 3-D gradient vectors $\mathbf{g}_i = dB_i/d\mathbf{X}$, what is required is to find the vector $\mathbf{x}$ of unit length which minimizes the sum of projection lengths $\mathbf{x} \cdot \mathbf{g}_i$. This is done by finding the eigenvector for the least eigenvalue of the matrix

$$\mathbf{M} = \sum_{i=1}^{N}(\mathbf{g}_i\mathbf{g}_i^{\mathsf{T}}) \qquad (8)$$

where $(\mathbf{g}_i\mathbf{g}_i^{\mathsf{T}})$ is the 3×3 outer product of the vector $\mathbf{g}_i$ with itself.

If the third eigenvalue is not small relative to the first and second, there is no vector which is a satisfactory approximation to the normal of the inputs. In that case, the 3-D location selecting these image measurements must not be a part of a consistent object; we carve the location away.

At surface discontinuities, the image gradient is due to a boundary edge for the object. The 3-D gradient describes the tangent plane passing through the camera center and the limb of the object. The surface at the limb may be a sharp corner, or it may be a smooth surface. In either case, the tangent plane containing the camera center and the edge will be consistent with any iso-intensity direction on the surface; the 3-D gradient is normal to *all* directions on the surface. If the edge is due to a sharp corner, the same edge may be visible in several views and the iso-intensity contour can be determined precisely. In the case of a smooth surface edge, the iso-intensity direction will have to be determined through texture edges on the surface seen from other viewpoints. A smooth surface with a smooth texture will not have a single iso-intensity direction, but at the limb of the object it can be limited to the tangent plane.

5 The normal motion field in three dimensions

5.1 Normal flow

The motion constraint equation is the mathematical formulation of the statement that the brightness of a point in the scene remains constant for small motions. Formally,

$$\frac{\partial I}{\partial x}\frac{dx}{dt} + \frac{\partial I}{\partial y}\frac{dy}{dt} + \frac{\partial I}{\partial t} = 0. \qquad (9)$$

This equation relates the component of the optical flow parallel to the image gradient to the time derivative of the intensity. Since this optical flow component is normal to the edges of the image, it is termed *normal flow*.

The tangential component of the optical flow cannot be determined directly without additional assumptions. These extra constraints introduce biases into the computation of flow [5] [6]. Flow smoothness constraints may be marginally acceptable in applications where the camera is moving in a rigid environment. However, with a stationary camera and a non-rigid scene, we expect regions of zero flow to be near regions of quite high flow values. Refusing to compute flow near discontinuities will leave us without the most informative parts of the scene. Rather than accepting additional biases or loss of measurements in order to compute a quantity which is not well-defined in the image, we choose to work with the normal component of the flow.

5.2 Three dimensions

Just as in the 2-D case, we start from the premise that motion along the iso-brightness contour cannot be measured locally. This is simply the restatement of the aperture problem in the 3-D framework. The case is not so bad in three dimensions, as we can still determine motion in two out of three principal axes.

The normal motion on an iso-brightness contour is by definition limited to the plane of normals to the contour at a given point. The component of the motion parallel to the contour cannot be measured directly, and doesn't concern us here. Each measurement of normal flow in the image set defines a line in the image, called the normal flow constraint line. Both the optical flow and the projection of the 3-D normal flow must lie along this constraint line. This line, together with the camera center, defines a constraint plane through scene space. The constraint planes from the various views intersect the plane of normals in a set of lines, all of which intersect in the single point defining the 3-D normal flow. If the lines fail to intersect, this indicates that the normal flow measurements in the images are inconsistent.

In practice, there are errors in normal flow measurements, and in gradient measurements. We need to take these errors into account when designing the algorithm which determines the consistency and value of the 3-D normal flow. When more than two normal flow measurements are available, the problem of finding the best intersection point is an optimization problem. The point which minimizes the sum of squared distances from the constraint planes (including the plane of edge normals) is found, while the error measurement determines whether the selected point is sufficiently consistent to use as the 3-D normal flow.

We find the constraint plane for a given image point $\mathbf{x}$, normal flow $[n_1, n_2]$, and the iso-intensity direction defined in the previous section. This plane is defined in homogeneous coordinates by the camera center $\mathbf{C}$, the direction vector $\mathbf{D} = \mathbf{P}^{-1} (x_1 + n_1, x_2 + n_2, 1)^{\mathsf{T}}$ —as in (2)— and the iso-intensity direction unit vector $\mathbf{W}$. Using homogeneous coordinates for the point $\mathbf{C}$, and representing the vectors $\mathbf{D}$ and $\mathbf{W}$ by the point at infinity in the vector direction, the plane $\mathbf{p}$ containing these (homogeneous) points is given by the determinant

$$
\mathbf{p} = \begin{vmatrix} \mathbf{e}_1 & \mathbf{e}_2 & \mathbf{e}_3 & \mathbf{e}_4 \\ C_1 & C_2 & C_3 & 1 \\ D_1 & D_2 & D_3 & 0 \\ W_1 & W_2 & W_3 & 0 \end{vmatrix} \tag{10}
$$

$$
= [(\mathbf{D} \times \mathbf{W}), -[\mathbf{DWC}]]. \tag{11}
$$

Likewise, the plane of normals to $\mathbf{W}$ through $\mathbf{X}$ combines the other two eigenvectors of the matrix $\mathbf{M}$ of (8), $\mathbf{U}$ and $\mathbf{V}$. Noting that $\mathbf{W} = \pm \mathbf{U} \times \mathbf{V}$, the homogeneous coordinate of this plane is

$$
\mathbf{q} = [(\mathbf{U} \times \mathbf{V}), -[\mathbf{UVX}]] \tag{12}
$$

$$
\equiv [\mathbf{W}, -\mathbf{W} \cdot \mathbf{X}]. \tag{13}
$$

To find the best candidate for the intersection of these planes, we need to weight the planes equally in Euclidean space. This involves scaling the homogeneous coordinates $\mathbf{p}$ by $1/\| [p_1, p_2, p_3] \|_2$. Scaled this way, the homogeneous coordinate can be viewed as a unit vector in the direction normal to the plane, and the negative of the distance from the origin to the plane along this vector. The point $\mathbf{X}'$ is then constrained to have $X_4' = 1$ and is the least squares solution minimizing the error function

$$
E = \sum_{i=1}^{i<N} (\mathbf{p}_i \cdot \mathbf{X}')^2. \tag{14}
$$

This gives three constraint equations for $k \in \{1, 2, 3\}$ of

the form

$$
-\sum_{i=1}^{N} p_{i,k} p_{i,4} = \sum_{j=1}^{3} X_j' \sum_{i=1}^{N} p_{i,k} p_{i,j}. \tag{15}
$$

The normal motion vector is then $\mathbf{N} = \mathbf{X}' - \mathbf{X}$. In places where the error measure of (14) is excessive, the normal flow values can be deemed inconsistent, and the 3-D point $\mathbf{X}$ can be carved away. Alternately, the reprojection of the vector $\mathbf{N}$ into each of the images can be checked for consistency with the normal flow measurement in that image. The projection of $\mathbf{X}'$ should lie on (near) the motion constraint line.

6 Results

Here we present some results of the algorithm. Input data was obtained from sixteen cameras widely separated around the room, with a person walking through the scene. Cameras were calibrated using images of a known calibration object, and images were corrected for radial distortions. These reconstructions used images of 320×240 pixels. Four of the input images are shown in Figure 1; Figure 1(a) and (b) are two two raw data images, (c) is an image of the background, and (d) is the foreground silhouette.

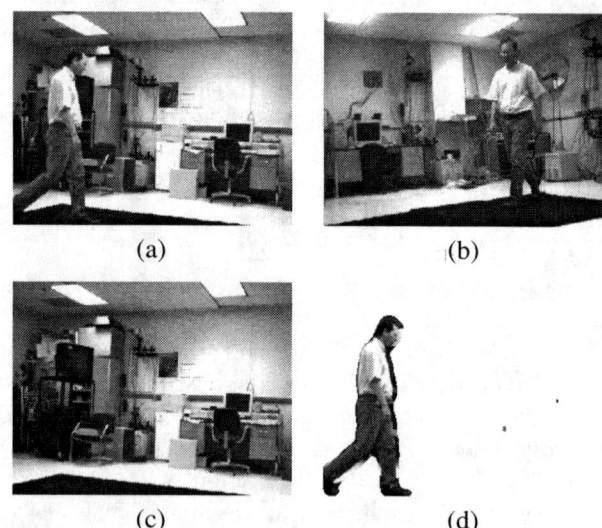

(a)   (b)

(c)   (d)

Figure 1. Input Data

In Figure 2, images (a) and (b) show two views of a moving person, after the data structure has been initialized with the silhouette intersection from twenty views. The foreground/background separation procedure purposely favors the foreground, as any missing foreground in a view will generate holes in the initial volume. This also adds a shell around the volume which needs to be carved away, along with any concavities in the moving human. Figures 2(c) and (d) show a depth map of the scene before and after the

86

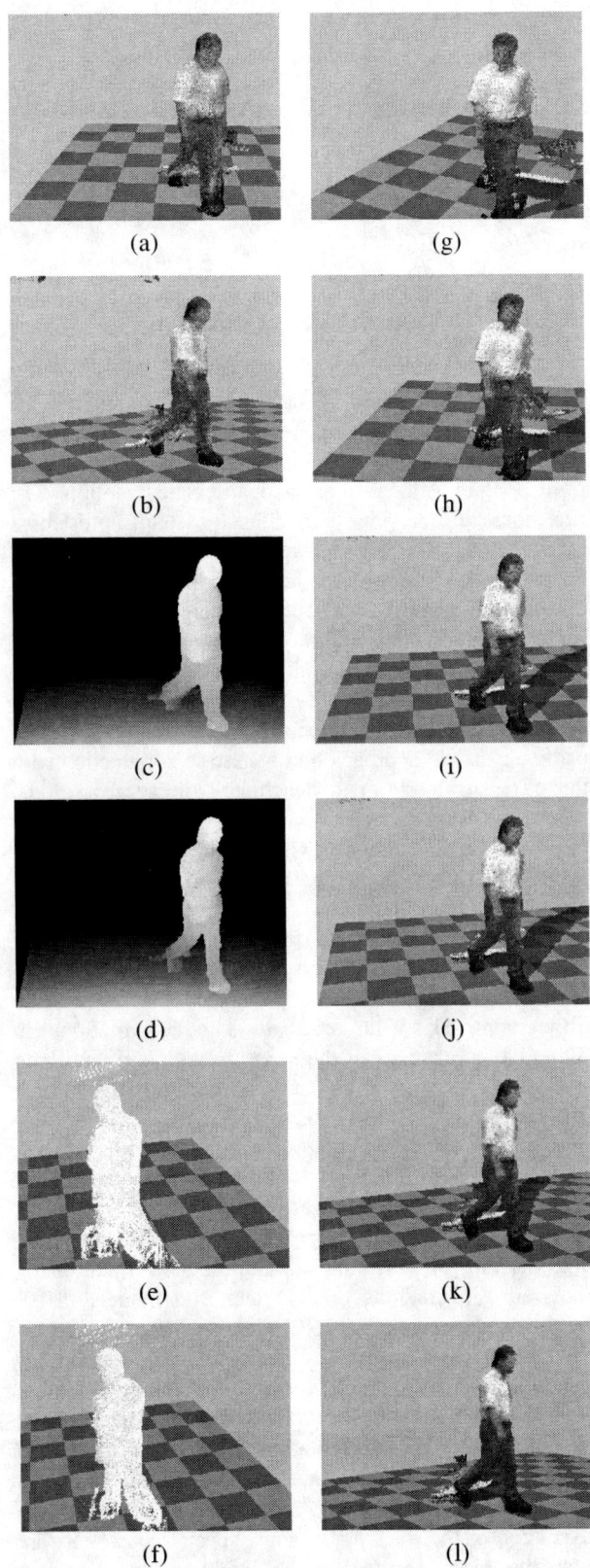

Figure 2. Results

ray carving. Figures 2(e) and (f) show the agreement in the iso-intensity direction before and after the carving. Light colors represent maximum agreement, darker colors indicate inconsistency. Figures 2(g) through (l) show five views of the scene from viewpoints between the cameras. A virtual floor and shadow are added to the scene, as a simple example of the effects available using full 3-D structure.

References

[1] P. Baker and Y. Aloimonos. Complete calibration of a multi-camera network. In *Proc. IEEE Workshop on Omnidirectional Vision*, pages 134–141, Hilton Head Island, SC, 2000. IEEE Computer Society.

[2] W. B. Culbertson, T. Malzbender, and G. Slabaugh. Generalized voxel coloring. In *Vision Algorithms: Theory and Practice*, Corfu, Greece, 1999. IEEE.

[3] L. Davis, E. Borovikov, R. Cutler, D. Harwood, and T. Horprasert. Multi-perspective analysis of human action. In *Proc. of Third International Workshop on Cooperative Distributed Vision*, Kyoto, Japan, 1999.

[4] O. D. Faugeras. *Three-Dimensional Computer Vision*. MIT Press, Cambridge, MA, 1992.

[5] C. Fermüller, R. Pless, and Y. Aloimonos. Statistical biases in optic flow. In *Proc. IEEE Conference on Computer Vision and Pattern Recognition*, volume 1, pages 561–566, 1999.

[6] C. Fermüller, R. Pless, and Y. Aloimonos. The Ouchi illusion as an artifact of biased flow estimation. *Vision Research*, 40:77–96, 2000.

[7] K. N. Kutulakos and S. M. Seitz. What do *N* photographs tell us about 3D shape? Computer Science Technical Report 692, University of Rochester, 1998.

[8] P. Narayanan, P. Rander, and T. Kanade. Constructing virtual worlds using dense stereo. In *Proc. International Conference on Computer Vision*, pages 3–10, Bombay, 1998.

[9] H. Saito and T. Kanade. Shape reconstruction in projective grid space from large number of images. In *Proc. IEEE Conference on Computer Vision and Pattern Recognition*, volume 2, pages 49–54, 1999.

[10] S. M. Seitz and C. Dyer. Photorealistic scene reconstruction by voxel coloring. In *Proc. IEEE Conference on Computer Vision and Pattern Recognition*, pages 1067–1073, 1997.

[11] M. E. Spetsakis and J. Aloimonos. A unified theory of structure from motion. In *Proc. DARPA Image Understanding Workshop*, pages 271–283, 1990.

[12] S. Vedula, S. Baker, P. Rander, R. collins, and T. Kanade. Three-dimensional scene flow. In *Proc. International Conference on Computer Vision*, Corfu, Greece, 1999.

[13] Z. Zhang, O. D. Faugeras, and N. Ayache. Analysis of a sequence of stereo scenes containing multiple moving objects using rigidity constraints. In *Proc. Second International Conference on Computer Vision*, pages 177–186, 1988.

Tracking Multiple Objects in the Presence of Articulated and Occluded Motion

Shiloh L. Dockstader and A. Murat Tekalp
Department of Electrical and Computer Engineering
University of Rochester, Rochester, NY 14627 USA
URL: http://www.ece.rochester.edu/~dockstad/research/

Abstract

This paper presents a novel approach to the tracking of multiple articulate objects in the presence of occlusion in moderately complex scenes. Most conventional tracking algorithms work well when only one object is tracked at a time. However, when multiple objects must be tracked simultaneously, significant computation is often introduced in order to handle occlusion and to calculate the appropriate region correspondence between successive frames. We introduce a near real-time solution to this problem by using a probabilistic mixing of low-level features and components. The algorithm mixes coarse motion estimates, change detection information, and unobservable predictions to create accurate trajectories of moving objects. We implement this multi-feature mixing strategy within the context of a video surveillance system using a modified Kalman filtering mechanism. Experimental results demonstrate the efficacy of the proposed tracking and surveillance system.

1. Introduction

Object tracking is an important task for a variety of applications within the areas of digital video processing and computer vision. Some of the more prevalent applications include autonomous vehicular navigation, robotic control, motion-based recognition, and video scene surveillance. Several of the aforementioned applications are likely to require the tracking of multiple objects in a real-time, or near real-time, manner. Unfortunately, there typically exists a compromise between the achieving of real-time tracking performance and the constructing of a tracking algorithm that is robust to multiple occlusions. In order to formulate a tractable problem, we limit our application of multiple object tracking to the field of video surveillance.

We present the framework for an autonomous video surveillance system that successfully monitors a scene for moving people. The system must be capable of operating in real-time on standard PC hardware and be robust to the presence of multiple people, people-people and people-environment occlusions, moderate lighting changes, non-rigid contour deformations, and articulated motion. The first phase of our video surveillance system represents a tracking module which records the precise trajectories of all moving video objects; the second phase develops a database of images which contains a single facial snapshot for every human subject in the video sequence. To address the former problem we introduce a novel technique that is capable of tracking multiple occluding objects in complex scenes. The latter issue of face recognition is implemented using a classical approach and is used in conjunction with the tracking algorithm to benchmark the success of the video surveillance system.

A strong foundation in video object tracking is likely to include some combination of change detection, motion estimation, spatio-temporal filtering, and the like. For real-time performance and pixel accuracy along object boundaries, change detection plays an important role. In particular, we consider the use of dynamic reference frame differencing [1-4] in order to handle variations in illumination, changes in the scene background, and long periods of exiguous object motion. Although frame differencing is useful in detecting motion, it is inherently incapable of predicting non-translational object movement, thus limiting its utility in the tracking of articulated and occluded object motion.

More sophisticated approaches to the segmentation and tracking of video object motion have traditionally required combinations of 2-D and 3-D models [5-10] mesh-based constraints [11,12], deformable contours [13-15], and dense motion estimation. These approaches generally provide reliable tools for tracking, but often require extensive processing, multi-view implementations, or camera calibration information. Extending model-based tracking to real-time scenarios has proven to be a challenging, yet tractable, problem. To quicken the execution of the tracking systems, special constraints are typically placed on either the hardware (e.g., custom hardware) or the image sequences (e.g., sequences with monochromatic background regions) [8-10].

A variety of object tracking techniques that extend beyond the aforementioned categories have shown fruitful results on real image sequences. Brémond and Thonnat [16] present a method for tracking multiple non-rigid objects in video sequences. They define a target model by using the midpoints along each of the four boundaries as well as the center point of a bounding box that may contain one or more moving objects. McKenna *et al.* [17] present an algorithm that is capable of tracking objects in real-time based solely on the use of adaptive Gaussian mixture models. These mixture models describe the composition of a color object in hue-saturation space. The model adapts the color distributions through successive frames and produces fairly accurate results. Since the algorithm relies only on color, however, it is easily disrupted in the presence of object occlusion. In such cases, the Gaussian mixtures tend to deviate from the desired target and begin training on inaccurate data, which causes the tracking and recognition to fail.

One of the more popular and successful techniques used for the tracking of video objects is Kalman filtering [1,3,4,13,18]. Traditional Kalman filtering models occlusion as Gaussian observation noise. This assumption works well in many cases, but can be improved upon when independent Kalman filters begin to interact with each other. We introduce (i) a novel implementation of the Kalman filter that uses change detection to generate observations and coarse motion estimates to generate predictions and (ii) a new probabilistic coupling for multiple Kalman filters. We take advantage of the fact that the observations and predictions for interfering Kalman filters are no longer independent. These probabilistic modifications to the Kalman filtering configuration are integrated as confidence measures that take the form of symmetric weighting matrices. The result is a significant improvement in the accuracy of tracking multiple occluded and articulated objects.

2. System overview

To allow the surveillance system to achieve real-time performance, we impose a number of restrictions on the video sequence: (i) the sequence is captured using a stationary camera with a fixed focal length; (ii) if the sequence contains moving people, they are moving in an upright fashion; and (iii) the video camera is positioned such that the 3-D point intersection between the optical axis and the ground plane projects onto the image plane. The final assumption provides a reasonable foundation for occlusion reasoning while using a stationary camera [4]. We place no restrictions on occlusion (save that it should

not be complete), object motion, or the conditions of the initial video frame. Since the proposed technique does not attempt to introduce a novel approach for finding faces, we simply assume that if a face exists it will be found in the upper $\frac{1}{6}^{th}$ of a bounding box. This ratio is based on actual training data [1] and is surprisingly accurate for the majority of moving people.

Before using the system in the real-time mode, it is first necessary to determine a number of operational parameters; λ_D, λ_M, λ_T. These parameters represent a normalized distance threshold (λ_D), a normalized motion threshold (λ_M), and a time-frame threshold (λ_T). In order to optimize the selection of these three parameters, we choose a training sequence and an optimization criterion. The training sequence consists of several thousand frames of moving people in the presence of illumination changes, background noise, a variety of facial poses, and the like. The optimization criteria is based on receiving operator characteristic (ROC) curve accuracy maximization [19], where accuracy is defined by

$$Accuracy(\lambda_D, \lambda_M, \lambda_T) \triangleq \frac{TP + TN}{TP + FP + FN + TN} \quad (1)$$

and *TP* indicates the total number of regions said to contain facial pixels that actually contain facial pixels and *TN* specifies the total number of regions said to contain no facial pixels that actually contain no facial pixels. In a similar manner, *FP* denotes the number of regions said to contain facial pixels that actually do not while *FN* indicates the number of regions which contain facial pixels but which are missed by the system. The classification results are taken from the entire training sequence as a function of λ_D, λ_M, and λ_T. The goal is to maximize the function specified by (1) and to then use those values of λ_D, λ_M, and λ_T at which the global maximum is achieved for all subsequent video sequences.

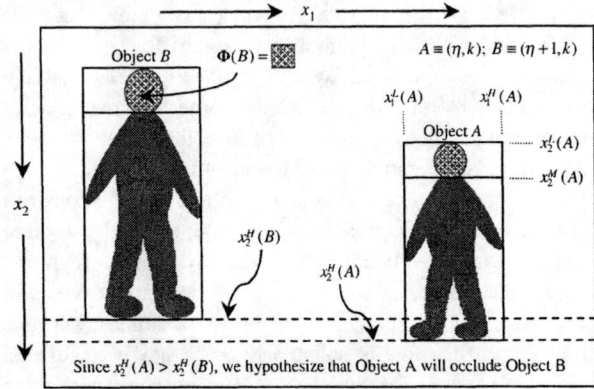

Figure 1. Bounding box modeling and notation.

When determining the system accuracy, we automatically declare a capture to be a failure (regardless

89

of facial content) if $x_1^L(\eta,k)$ or $x_1^H(\eta,k)$, as defined in Figure 1, deviate from their ideal locations by more than 25% of the ideal width of the bounding box or if $x_2^L(\eta,k)$ or $x_2^H(\eta,k)$ deviate by more than 5% of the ideal height of the bounding box. This measure is somewhat tedious to calculate for an entire video sequence, but necessary in order to determine an approximation to the quality of the tracking and surveillance performance. This measure also provides a reasonable balance between the need for an analysis of the robustness of the surveillance system and the need to benchmark the success of the novel tracking modifications. A pixel-accurate error for each frame is both burdensome to calculate and inappropriate, as the tracking algorithm does not necessarily claim to produce pixel-accurate results under arbitrary circumstances.

3. Algorithm theory

The following sections provide overviews of the motion detection and estimation, object discrimination, and facial capture while introducing a thorough treatment of the proposed object tracking concepts. For a more precise handling of the facial capture, motion detection and estimation, and object discrimination theory we refer the reader to [20].

3.1 Motion detection and estimation

Change detection constitutes the primary tool used for detecting and locating areas of potential motion. The proposed system uses the motion threshold, λ_M, in order to determine those areas in the current frame that significantly differ from the reference frame. The output is a binary map, $w_k(\mathbf{n})$.

We use dynamic frame differencing to perform change detection between the current frame and some temporally varying background image [reference frame]. The reference frame, $\mathbf{r}_k(\mathbf{n})$, updates any pixels that have remained relatively stationary in color for λ_T or more frames. The update does not apply to pixels that are known to belong to foreground objects; that is, this exception caters to foreground objects that enter the scene and then stop moving. Although the algorithm stops placing bounding boxes around such regions, the positions are held in memory to facilitate the tracking of other objects that may become occluded by these foreground areas. An object in the scene that suddenly leaves will disrupt the reference frame for up to λ_T frames, but then the reference frame is updated in those newly uncovered areas to account for the change. The proposed technique for extracting the background is reliable, robust, and very efficient.

The algorithm estimates motion correspondence based on optical flow over a sparse grid only in areas that have been flagged as changed regions from the reference frame. The coarse motion output, $y_k(\mathbf{n})$, is used to assist with the tracking of articulated moving objects, develop accurate predictions for the Kalman filter, and cluster the foreground regions into separate objects.

3.2 Object discrimination

Spatial clustering / object discrimination consists of a functional mapping from the change detection map, $w_k(\mathbf{n})$, to a scalar-valued segmentation map, $z_k(\mathbf{n})$. This mapping is implemented via four key steps: (i) region isolation, (ii) region merging, (iii) region filtering, and (iv) region splitting. Region isolation uses two passes through $w_k(\mathbf{n})$ to label connected components. The first pass creates an equivalence table for the change detection map, calculates the size and average motion of each isolated region, and determines a precise bounding box for each isolated region.

Region merging is a process that takes multiple areas of $z_k(\mathbf{n})$ and reclassifies them as a single connected region. The merging step provides the clustering routine with a means for handling noise and foreground pixels that exhibit colors similar to the background regions that they occlude. Region filtering is a simple mechanism that allows the algorithm to maintain a level of robustness to noise and unstable motion in the video sequences. The filtering is based on both the size of isolated regions as well as the distance of these regions to previously known points of motion (e.g., the distance from image borders, where new objects are expected to appear, and the distance from previously tracked objects).

The final step of the clustering routine splits regions by combining the tracking results from previous frames with motion estimates generated from the current frame. Allowing regions to split provides the Kalman filtering process with a set of observations for occluding and occluded objects. The observations are based only on a translational model, however, and are similar to those that might be obtained using only change detection. It will be the function of the Kalman filter to weight these observations against the [possibly] more accurate predictions generated from the coarse motion estimates.

3.3 Object tracking

The process of tracking objects in video sequences becomes increasingly difficult as the number of objects increases. Although, in itself, the tracking of deformable and articulated objects is a difficult problem, this is compounded by occlusion. Addressing occlusion in real-time sequences further adds to the dilemma by limiting the types of techniques that can feasibly be applied to the analysis of occluded regions. Kalman filtering, depending on the number of states, can be a very fast and effective technique for tracking objects in the presence of occlusion. Since our approach to tracking is driven by the application of video surveillance, we model a Kalman filtering technique with a state vector based on the bounding boxes for each of the tracked objects. In particular, the state vector takes the form

$$\mathbf{s}_\eta[k] \equiv \begin{bmatrix} x_1^L(\eta,k) & x_1^H(\eta,k) & x_2^L(\eta,k) & x_2^H(\eta,k) \end{bmatrix}^T, \quad (2)$$

where, as demonstrated in Figure 1, $x_1^L(\eta,k)$, $x_1^H(\eta,k)$, $x_2^L(\eta,k)$, $x_2^H(\eta,k)$ represent the left, right, upper, and lower sides of the ηth bounding box in the kth frame.

A traditional Kalman filter models occlusion as Gaussian observation/prediction noise. This works moderately well when some unknown object in the environment occludes a tracked subject (e.g., when a tracked subject walks behind a tree or pillar). In principle, some quality metric that measures the accuracy of the tracking is recursively used to modify the noise covariance matrices of the filter – the result is a shift towards a pure predictive estimator. Typically, an approach using correlation is used in conjunction with change detection in order to develop an accurate quality metric for the Kalman filter. Correlation works well for non-deformable and rigid locomotion, but implementing a deformable correlator for tracking articulated objects, such as human body parts, can be computationally burdensome for real-time applications. We address the aforementioned difficulty by introducing a temporally varying weight matrix, $\Xi_\eta[k]$, to the Kalman filter's minimum prediction mean-square error (MSE) formulation.

The first step in Kalman filtering involves the estimation of a state prediction for the current frame, k, according to

$$\hat{\mathbf{s}}_\eta[k \mid k-1] = \mathbf{A}_\eta[k]\hat{\mathbf{s}}_\eta[k-1 \mid k-1] \quad (3)$$

where $\hat{\mathbf{s}}_\eta[k \mid k-1]$ is the state prediction for the current frame, $\mathbf{A}_\eta[k]$ is the time-varying state transition matrix, and $\hat{\mathbf{s}}_\eta[k-1 \mid k-1]$ is the state estimate from the previous frame for the ηth object. To develop the most accurate prediction for the current frame, we construct $\mathbf{A}_\eta[k]$ using coarse motion estimates that are directed from the previous frame to the current frame which are known to have originated within the foreground region associated with the ηth object in that frame. The use of coarse motion estimates is accurate enough to allow each of the four bounding box edges associated with a single region to move independently, if necessary. Simple change detection does not afford this convenience or accuracy.

After creating a state prediction for the current frame, one must develop a measure of confidence for that prediction. The minimum prediction MSE matrix, $\mathbf{M}_\eta[k \mid k-1]$, is used for this purpose. The traditional formulation for the Kalman filter uses the updated MSE matrix from the previous frame by predicting a new value and adding an additional Gaussian noise covariance that correlates with the accuracy of the predicted data according to

$$\mathbf{M}_\eta[k \mid k-1] = \mathbf{\Gamma}_\eta[k] + \mathbf{Q}_\eta[k], \quad (4)$$

where

$$\mathbf{\Gamma}_\eta[k] = \mathbf{A}_\eta[k]\mathbf{M}_\eta[k-1 \mid k-1]\mathbf{A}_\eta^T[k]. \quad (5)$$

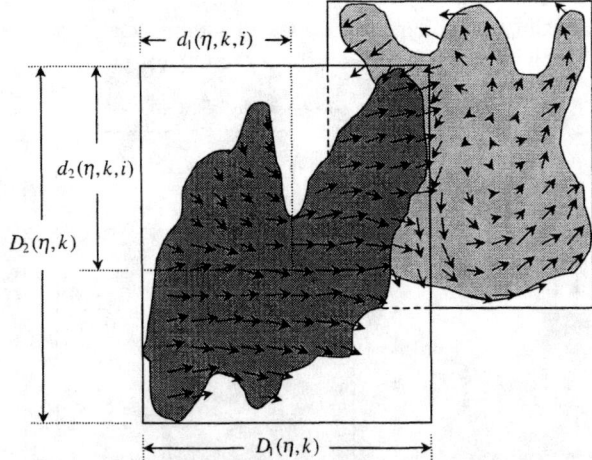

Figure 2. Kalman prediction mechanism. We illustrate two objects, each shown with its corrected bounding box (from the k-1th frame). We use $d_1(\eta,k,i)$ and $d_2(\eta,k,i)$ to indicate the origin of the ith motion vector within the ηth object.

The matrix, $\mathbf{Q}_\eta[k]$, is recursively modified over time to account for the deviations between the predictions and corrections of the state vector estimates. Since the error of the predictions is not always Gaussian in nature and the appropriate magnitude of $\mathbf{Q}_\eta[k]$ is not necessarily known *a priori*, we find it possible to improve the accuracy of the minimum prediction MSE matrix by introducing a weight matrix, $\Xi_\eta[k]$, that directly correlates with the technique used to generate the state transition matrix, $\mathbf{A}_\eta[k]$, from the coarse motion estimates. In particular, we consider modifying (5) so that

$$\mathbf{\Gamma}_\eta[k] = \mathbf{A}_\eta[k]\Xi_\eta[k]\mathbf{M}_\eta[k-1 \mid k-1]\Xi_\eta^T[k]\mathbf{A}_\eta^T[k] \quad (6)$$

where

$$\Xi_\eta[k] \triangleq diag\left\{ \begin{bmatrix} \frac{2}{N_\eta^k}\sum_{i=1}^{N_\eta^k}\frac{d_1(\eta,k,i)}{D_1(\eta,k)} & \frac{2}{N_\eta^k}\sum_{i=1}^{N_\eta^k}\frac{1-d_1(\eta,k,i)}{D_1(\eta,k)} & \cdots \\ \cdots & \frac{2}{N_\eta^k}\sum_{i=1}^{N_\eta^k}\frac{d_2(\eta,k,i)}{D_2(\eta,k)} & \frac{2}{N_\eta^k}\sum_{i=1}^{N_\eta^k}\frac{1-d_2(\eta,k,i)}{D_2(\eta,k)} \end{bmatrix} \right\}, (7)$$

and $d_1(\eta,k,i)$, $d_2(\eta,k,i)$, $D_1(\eta,k)$, and $D_2(\eta,k)$ are as shown in Figure 2 and N_η^k indicates the total number of coarse motion estimates that originated within the foreground of the ηth object in the k-1th frame. In essence, this approach allows the algorithm to modify the strength of the prediction MSE matrix according to the distance of the coarse motion estimates from the position of the states for the Kalman filter. For instance, a state prediction based on coarse motion estimates that are very close in Euclidean distance to the state under consideration will produce a lower entry along the associated diagonal element in (7). Alternatively, a prediction based on motion vectors that are situated further from the state position should produce a higher entry for that state in the weight matrix.

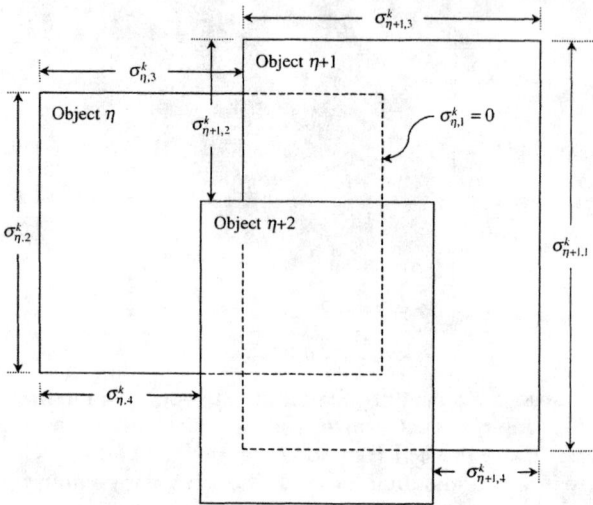

Figure 3. Kalman observation modeling. Object layering in the presence of occlusion explicitly utilizes $\alpha_{\eta,t}^{k} \triangleq \sigma_{\eta,t}^{k} \{D_2(\eta,k)\}^{-1}\big|_{t\in\{1,2\}}$ **and** $\alpha_{\eta,t}^{k} \triangleq \sigma_{\eta,t}^{k} \{D_1(\eta,k)\}^{-1}\big|_{t\in\{3,4\}}$.

We assume that the coarse motion vectors, on average, will follow a uniform distribution over the support of a particular bounding box; when this condition is met precisely, we have $\boldsymbol{\Xi}_\eta[k] = \mathbf{I}$. This form suggests that the weight matrix has little or no effect when the observations for the Kalman filter closely match the corrected predictions. On the other hand, as the observations and predictions begin to deviate, the weight matrix either amplifies or attenuates those states in the prediction MSE matrix. This provides the Kalman filter with more precise control over the weight given to predictions during periods of occlusion by unknown foreground objects. The result is a confidence measure, directly correlated with the formation of $\mathbf{A}_\eta[k]$, which improves upon the accuracy of tracking occluded objects.

The next step of the linear Kalman filter involves the construction of the Kalman gain matrix, $\mathbf{K}_\eta[k]$. This matrix determines the magnitude of the correction necessary to produce the most accurate state estimate for a given instant in time. The operation relies on $\mathbf{M}_\eta[k \mid k-1]$ and a noise covariance matrix, $\mathbf{C}_\eta[k]$, which describes the distribution of the assumed Gaussian observation noise. The approach works well for the general case, where little or no information is available concerning potential obstructions in the scene of the video sequence. When tracking multiple objects, one conventionally assigns a different Kalman filter to each of the independently moving objects. However, as the independently moving objects begin to interact (e.g., one object occludes the other), the observations obtained for each of the moving objects are no longer independent. We suggest an improvement to the Kalman filtering formulation that explicitly utilizes occlusion information for the states of multiple Kalman filters. The improvement takes the form of a probabilistic confidence matrix, $\mathbf{W}_\eta[k]$, which effectively couples the operation of multiple, neighboring Kalman filters while introducing only a small amount of additional computation.

We construct $\mathbf{C}_\eta[k]$ in the traditional way by allowing the matrix to be representative of the time-varying differences between observations and corrected predictions. We then construct a Kalman gain matrix according to

$$\mathbf{K}_\eta[k] = \mathbf{M}_\eta[k \mid k-1]\mathbf{H}_\eta^T[k]\left(\boldsymbol{\Lambda}_\eta[k] + \boldsymbol{\Psi}_\eta[k]\right)^{-1} \quad (8)$$

where $\mathbf{H}_\eta[k]$ represents the linear observation matrix,

$$\boldsymbol{\Lambda}_\eta[k] = \mathbf{W}_\eta[k]\mathbf{C}_\eta[k]\mathbf{W}_\eta^T[k], \quad (9)$$

$$\boldsymbol{\Psi}_\eta[k] = \mathbf{H}_\eta[k]\mathbf{M}_\eta[k \mid k-1]\mathbf{H}_\eta^T[k], \quad (10)$$

and

$$\mathbf{W}_\eta[k] \triangleq \left(diag\left\{\left[\alpha_{\eta,1}^{k} \ \ \alpha_{\eta,2}^{k} \ \ \alpha_{\eta,3}^{k} \ \ \alpha_{\eta,4}^{k}\right]\right\}\right)^{-\frac{1}{2}} \quad (11)$$

where $\alpha_{\eta,i}^{k}$, $1 \le i \le 4$ is a probabilistic measure for the ηth box in the k-1th frame, as shown in Figure 3. As indicated by the figure, the diagonal elements of $\mathbf{W}_\eta[k]$ are inversely proportional to the visibility of the states associated with a particular object. In the limiting case where a single object is being tracked we have $\mathbf{W}_\eta[k] = \mathbf{I}$, thus suggesting that the proposed technique has no negative impacts when multiple objects are not being tracked simultaneously. Since the formation of $\mathbf{W}_\eta[k]$ depends on the previously corrected states of all neighboring objects, we introduce an inherent coupling of the Kalman gain matrices for all of the associated Kalman filters.

For the sake of completeness, we provide the remaining steps of the Kalman filtering procedure. These equations follow those of the standard Kalman filter and are indicated by

$$\hat{\mathbf{s}}_\eta[k \mid k] = \hat{\mathbf{s}}_\eta[k \mid k-1] + \mathbf{K}_\eta[k]\boldsymbol{\Theta}_\eta[k], \quad (12)$$

where

$$\boldsymbol{\Theta}_\eta[k] = \mathbf{x}_\eta[k] - \mathbf{H}_\eta[k]\hat{\mathbf{s}}_\eta[k \mid k-1] \quad (13)$$

and

$$\mathbf{M}_\eta[k \mid k] = \left(\mathbf{I} - \mathbf{K}_\eta[k]\mathbf{H}_\eta[k]\right)\mathbf{M}_\eta[k \mid k-1]. \quad (14)$$

Equations (12) and (14) represent the state correction based on the Kalman gain matrix and the update of the minimum MSE noise covariance matrix, respectively, where $\mathbf{x}_\eta[k]$ indicates observations for a particular object.

3.4 Facial capture

To perform face detection, we calculate chrominance histograms using only those pixels hypothesized as belonging to facial regions (i.e., pixels which exist as a subset of $\Phi(\eta,k)$, as previously presented in Figure 1). The calculated histograms are then compared to model histograms that are initially generated using annotated training data. As the person is tracked through the scene, the upper $\frac{1}{6}$ th of the bounding box is tested for a closer match to the model chrominance data. This ensures that only the *best* facial snapshot for a particular person is extracted from the sequence. This approach also requires that interactions between multiple moving persons not disrupt the tracking process (i.e., tracking switches from one person to another during occlusion).

4. Experimental results

The theory outlined above has been successfully tested on a number of sequences – we present the results on three of these sequences: *EKRL*, *VNJ*, and *Group*. The *Group* sequence consists of a number of individuals walking in an outdoor environment. This particular sequence suffers from noise and low resolution and includes a number of moving people in the first frame. The *VNJ* sequence exhibits less noise than the *Group* sequence, but contains significant levels of occlusion. A careful inspection of the sequence also shows variations in illumination due to moving clouds – this point is easily seen in the shadows of those moving in the scene. Finally, the *EKRL* sequence contains a number of manually imposed obstructions, very high levels of occlusion (due to both the environment and to other moving people), and noisy background regions. This sequence demonstrates the effects of introducing a foreign object (i.e., a jacket) to and removing a native object (i.e., a laptop case) from the scene. Since outdoor imaging was not favorable at the time of writing, we also simulate the effects of moderate wind levels on trees and plants. We encourage the reader to visit our website in order to see the tracking results for these, and other, video sequences in full.

We summarize the testing results for each of the video sequences in Table 1. For each sequence we provide a summary of the system accuracy (*SA*), as defined by (1), by analyzing the extracted values for *TP*, *TN*, *FP*, and *FN*. We also indicate the number of times the tracking

deviated from the *ideal* trajectory, according to our criterion outlined in the third paragraph of §2. This data is provided under two scenarios – in the first, we implement the system as proposed while in the second, we omit our suggested changes to the linear Kalman filter. In Table 1, we report on the tracking accuracy of both the proposed Kalman tracking system (denoted with a subscript *P*) and the classical Kalman tracking system (denoted with a subscript *B*), where the tracking accuracy is defined by

$$TA_i \triangleq \tau_i^1 \left(\tau_i^1 + \tau_i^0\right)^{-1}\Big|_{i \in \{B,P\}}. \quad (15)$$

In each frame of video we have some number of foreground objects tracked by the system (perhaps zero); τ_i^1 indicates the total number of correctly tracked objects over all frames. Similarly, τ_i^0 indicates the number of incorrectly tracked objects over all frames. An incorrect tracking occurs any time a foreground object is not followed with sufficient precision or any time a background object is erroneously tracked.

Table 1. Accuracy statistics.

	TN	TP	FN	FP	γ	η	τ_P^0	τ_P^1	τ_B^0	τ_B^1	DA	SA	TA_P	TA_B
Group	913	404	122	26	6	7	85	1482	127	1440	86%	90%	95%	92%
VNJ	69	259	26	14	3	3	25	344	84	285	100%	89%	93%	77%
EKRL	473	340	66	19	5	5	114	777	200	691	100%	91%	87%	77%

Since the aforementioned criteria for estimating the success of the surveillance system is rather conservative, we also use a more practical measure. Specifically, we include the number of facial snapshots actually added to the database and compare this number, γ, to our ideal metric, γ_i; this database accuracy ($DA \triangleq \gamma/\gamma_i$) is also reported in the table. The system should only add one image for each person, but no images for people who aren't facing the camera. Several resulting frames from each of the three sequences are presented in Figure 4; faces actually extracted from the system are shown in Figure 5.

Although the accuracy of the face detection leaves much to be desired, that of the object tracking is quite acceptable. As indicated in Table 1, the proposed tracking consistently outperforms that using only a classical Kalman filtering approach. In most cases, the tracking deviates any time there are no reliable observations for a bounding box edge and the underlying object is accelerating while being occluded. A pragmatic solution to this problem is likely to require a model-based tracking architecture. It should be noted that when the object tracking becomes irrevocably corrupted, we employ no heuristics or *ad hoc* strategies to help rectify the correspondence problems. In such a scenario, the tracking for the corrupted object yields failing results until it exits the scene. This fact triggers some of the large differences in success seen between the proposed tracking and that utilizing only a classical Kalman filter.

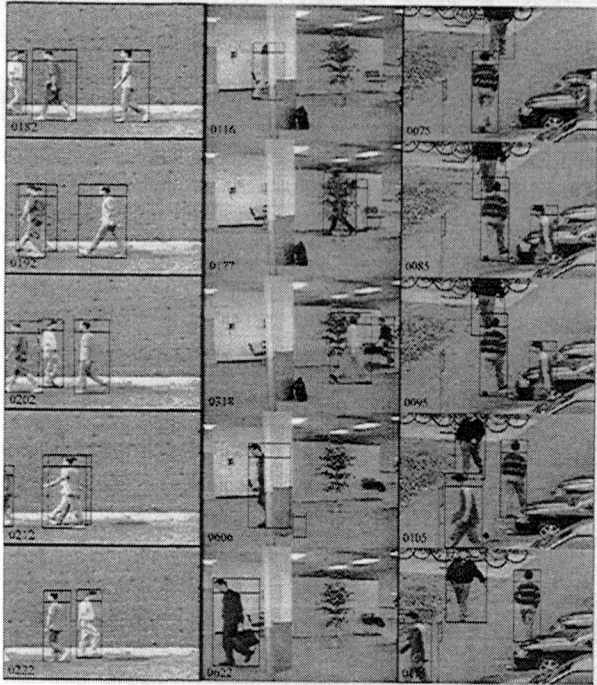

Figure 4. Tracking results. The first column includes five frames taken from the *VNJ* sequence, the second column is representative of the *EKRL* sequence, and the third column includes frames from the *Group* sequence.

Figure 5. Captured facial data.

The three image sequences were processed using a 330 MHz Sun Microsystems Ultra 10 workstation with Elite3D graphics. The amount of time required to process the data, after training, is furnished in Table 2 for each of the sequences. The speed of the algorithm is partly due to the conditional computations (i.e., on the presence of anticipated motion) in several modules. However, this property suggests that as the number of objects in the scene increases, so does the average processing time per frame. In our sequences, we found the effects to be acceptable, yielding an average processing rate of approximately 10 frames/sec. A practical implementation of this system would need to anticipate or assume a maximum number of foreground objects in order to prevent the dropping of video frames (a consequence which could adversely affect the tracking performance).

Table 2. Average processing times.

Image Sequence	Video Length (Frames)	Video Length (Seconds)	Total Processing Time (Seconds)	Average Time Per Frame
EKRL	676	22.5 s	70 s	0.104 s
VNJ	287	9.6 s	37 s	0.129 s
Group	1000	33.3 s	81 s	0.081 s
Total	1963	65.4 s	188 s	0.096 s

5. Conclusions and future work

We have shown how coarse motion estimation and dynamic reference frame differencing can be used in conjunction with a simple bounding box model to achieve real-time object tracking and recognition. The proposed technique introduces novel modifications to the standard Kalman filtering procedure to achieve robust tracking of multiple articulated and occluded moving objects. The approach is founded upon the use of change detection to provide pixel-accurate observations of non-occluded regions and the use of coarse motion estimation to develop sufficiently accurate predictions for partially occluded and/or articulated regions.

We introduce an explicit measure of confidence for the Kalman state predictions that is inversely proportional to the Euclidean distance between the positions of the motion vectors and the corresponding state locations. Additional Kalman filters are introduced for each new independently moving object in the scene. We take advantage of the fact that as independently moving objects interact (e.g., one object occludes another), the corresponding state predictions and observations for the multiple objects no longer remain independent. To that end, we introduce a probabilistic weight matrix that is computed based on the occlusion information extracted from the multiple moving objects. This weight matrix introduces an inherent coupling of the previously independent Kalman filters. These new modifications to the Kalman filter provide explicit modeling of information that is specific to the scene and to the motion of multiple objects that is not necessarily captured by the implicit recursions of the noise covariance matrices. The result is a noticeable improvement in the accuracy of tracking multiple occluded objects, with only a minor increase in computational complexity.

The proposed object tracking contribution is presented within the framework of an autonomous video surveillance system. The system takes advantage of the tracking accuracy and real-time performance by maintaining a reliable trajectory for each of the moving objects in the scene. To enhance the practicality of the surveillance system we suggest a simplistic technique that combines the tracking accuracy with a color-based mechanism for detecting faces of moving persons. The accuracy of the facial capture is used to benchmark the success of the tracking and recognition system.

Future work in this area intends to address potential improvements to both the tracking model and the recognition mechanism. Although not a focus of this paper, future endeavors will address more sophisticated methods of recognizing and tracking articulated and self-occluding objects in motion. We are currently addressing the possibility of extending the tracking from a bounding box model to more highly parameterized contours and surfaces. Ultimately, it is anticipated that the theory set forth in this paper will also be applied to improving the simultaneous tracking of individual body parts from multiple persons in occlusion.

Acknowledgments

We acknowledge Eastman Kodak Company for their financial support and Drs. Majid Rabbani and Gabriel Fielding for their many helpful suggestions and comments.

References

[1] O. Munkelt and H. Kirchner, "STABIL: A system for monitoring persons in image sequences," *Proc. of SPIE*, vol. 2666, pp. 163-179, 1996.

[2] S. Brofferio, L. Carnimeo, D. Comunale, and G. Mastronardi, "A background updating algorithm for moving object scenes," *Proc. of the Int. Workshop on Time-Varying Image Processing and Moving Object Recognition 2*, Florence, Italy, 1989, pp. 297-307.

[3] G. L. Foresti, "Object Recognition and Tracking for Remote Video Surveillance," *IEEE Trans. on Circuits and Systems for Video Technology*, vol. 9, no. 7, pp. 1045-1062, October 1999.

[4] D. Koller, J. Weber, and J. Malik, "Robust multiple car tracking with occlusion reasoning," *Proc. of the Eur. Conf. on Computer Vision*, Stockholm, Sweden, 2-6 May 1994, pp. 189-196.

[5] J. M. Rehg and T. Kanade, "Model-based tracking of self-occluding articulated objects," *Proc. of the Int. Conf. on Computer Vision*, Cambridge, MA, 20-23 June 1995, pp. 618-623.

[6] I. A. Kakadiaris and D. Metaxas, "Model-based estimation of 3D human motion with occlusion based on active multi-viewpoint selection," *Proc. of the Conf. on Computer Vision and Pattern Recognition*, San Francisco, CA, 18-20 June 1996, pp. 81-87.

[7] D. M. Gavrila and L. S. Davis, "Model-based tracking of humans in action: A multi-view approach," *Proc. of the Conf. on Computer Vision and Pattern Recognition*, San Francisco, CA, 18-20 June 1996, pp. 73-80.

[8] C.-C. Lien and C.-L. Huang, "Model-Based Articulated Hand Motion Tracking for Gesture Recognition," *Image and Vision Computing*, vol. 16, pp. 121-134, February 1998.

[9] T. Starner, J. Weaver, and A. Pentland, "Real-Time American Sign Language Using Desk and Wearable Computer Based Video," *IEEE Trans. on Pattern Analysis and Machine Intelligence*, vol. 20, no. 12, pp. 1371-1375, December 1998.

[10] Z. Li and H. Wang, "Real-Time 3D Motion Tracking with Known Geometric Models," *Real-Time Imaging*, vol. 5, no. 3, pp. 167-187, June 1999.

[11] C. Toklu, A. T. Erdem, M. I. Sezan, and A. M. Tekalp, "Tracking motion and intensity variations using hierarchical 2-D mesh modeling for synthetic object transfiguration," *Graphical Models and Image Processing*, vol. 58, no. 6, pp. 553-573, November 1996.

[12] P. J. L. van Beek, A. M. Tekalp, N. Zhuang, I. Celasun, and M. Xia, "Hierarchical 2-D mesh representation, tracking, and compression for object-based video," *IEEE Trans. on Circuits and Systems for Video Technology*, vol. 9, no. 2, pp. 353-369, March 1999.

[13] N. Peterfreund, "Robust Tracking of Position and Velocity with Kalman Snakes," *IEEE Trans. on Pattern Analysis and Machine Intelligence*, vol. 21, no. 6, pp. 564-569, June 1999.

[14] F. Leymarie and M. Levine, "Tracking Deformable Objects in the Plane Using an Active Contour Model," *IEEE Trans. on Pattern Analysis and Machine Intelligence*, vol. 15, no. 6, pp. 617-634, June 1993.

[15] A. Blake and M. Isard, *Active Contours*, Heidelberg, Germany: Springer-Verlag, 1998.

[16] F. Brémond and M. Thonnat, "Tracking Multiple Nonrigid Objects in Video Sequences," *IEEE Trans. on Circuits and Systems for Video Technology*, vol. 8, no. 5, pp. 585-591, September 1998.

[17] S. J. McKenna, Y. Raja, and S. Gong, "Tracking Colour Objects Using Adaptive Mixture Models," *Image and Vision Computing*, vol. 17, no. 3-4, pp. 225-231, March 1999.

[18] D. Davies, P. Palmer, and M. Mirmehdi, "Detection and tracking of very small low contrast objects," *Proc. of the British Machine Vision Conf.*, University of Southampton, UK, 14-17 September 1998, pp. 599-608.

[19] C. J. D'Orsi, "Screening Mammography Pits Cost Against Quality," *Diagnostic Imaging*, vol. 63, pp. 73-76, 1994.

[20] S. L. Dockstader and A. M. Tekalp, "Real-time object tracking and human face detection in cluttered scenes," *Proc. of SPIE*, vol. 3974, pp. 957-968, 2000.

Session 6

Multiple Persons or Multiple Cameras II

Human Motion Tracking System Based on Skeleton and Surface Integration Model Using Pressure Sensors Distribution Bed

Tatsuya Harada, Tomomasa Sato and Taketoshi Mori

The University of Tokyo

7-3-1 Bunkyo-ku, Hongo, Tokyo 113-8656 JAPAN

{harada, tomo, tmori}@ics.t.u-tokyo.ac.jp

Abstract

In this paper, we propose a human motion tracking system based on a full body model and a pressure sensors distribution bed. The full body model consists of a skeleton and a surface model. BVH files are used as the skeleton model that describes a hierarchy of joints and links. Wavefront Object files are used as the surface model that describes geometry of the surface. The bed has 210 pressure sensors that are under the mattress. It can measure pressure distribution image of a lying person. The lying person's motion is tracked by considering potential energy, momentum and a difference between the measured pressure distribution image and a pressure distribution image that is calculated by the full body model. Experimental results reveal that the realized system can track not only horizontal motions such as opening and closing legs but also vertical motions such as raising the upper body.

1 Introduction

Body movements are very important for humans to live. Moving the body makes humans take adaptive behavior to the outside world. It is said that physical conditions and mental conditions are buried in the body movements, because humans often move their bodies when they are in good health, but move rarely their bodies when they are in bad health. Therefore it is thought that physical and mental conditions can be estimated by measuring the body movements.

One third or a quarter of life is spent for being in bed. Especially people who need care usually spend the whole day for being in bed. If their movements are measured and evaluated quantitatively, these data can be used for a health monitoring and evaluation of rehabilitation progress.

Static Charge Sensitive Bed (SCSB) [1] is famous for monitoring the body movements in bed. SCSB can measure respiration, a heart rate and twitch movements. Temperature sensors distribution bed [2] can measure gross movements such as body turns. However, these beds cannot measure articular motions. Supine and lateral postures [3], body parts' positions [4] and infant's status (e.g., crying and sleeping) [5] can be recognized by using a pressure sensors distribution bed. However these systems cannot track motions such as twisted motions of the upper body.

As a body movement measuring system that sensors are attached to the body, AMI made "Actigraph" is widely known. The size of the Actigraph is about as large as a wristwatch. It can count body movements. There is the system that can recognize postures using acceleration sensors [6]. Although these sensors become small, many sensors must be attached to the body to measure articular movements. Since attaching many sensors to the body restricts person's activities, it is difficult to measure unaffected body movements.

Many researches are contributed for body motion tracking by using video camera. O'Rourke [7] proposed the body motion tracking algorithm by using human model constraints about 20 years ago. Rehg [8] studied fingers and palm tracking based on the kinematics constraints. Wren [9] realized three-dimensional upper body motion tracking system by using "blobs". However these systems cannot track the subtle upper body's twisting motions, which seem same postures as video images. It is difficult for these systems to extract face and hand features because the body is lost of sight in a quilt.

In our research, we developed lying person's motion tracking system. This system realizes unrestraint sensing by using a pressure sensors distribution bed and realizes motion tracking such as raising and twisting the upper body and opening and closing legs thanks to a skeleton and a surface integration model.

2 Body Movement

2.1 Importance of Body Movement

Importance of body movement measuring is widely recognized. Although measuring body movements are used for many fields, only clinical application is mentioned in this section.

A pressure ulcer is inflammation of skin. The bad circulation of the blood causes pressure ulcers when a patient loses perception of the skin or cannot move his or her body. One of the methods for predicting the occurrence of pressure ulcers is the "Braden scale". In the Braden scale, measuring activity of the body movement is an important item for predicting pressure ulcers.

There is a high correlation between a sleep stage (e.g., REM, N-REM) and body movements. It is known that body movements increase at the REM sleep stage. The sleep stage can be estimated by measuring brain waves. However, for infants, since infant's brain is immature, the body movement is better indicator of the sleep stage than the brain waves.

There are "Ramsey score" and "Command scale" which measure a sedation level of anesthesia. Body movement reaction to a pain stimulus is one of the indicators of Ramsey score. Command scale evaluates accuracy of body movement reaction to doctor's command.

As mentioned above, body movements are used to extract meanings, which do not reflect vital signs. If body movements are measured systematically, it is thought that physical and mental conditions can be estimated.

2.2 Types of Body Movement

Body movements are produced by muscle activity. Muscle is divided into cardiac muscle, smooth muscle and skeletal muscle. The cardiac muscle moves a heart. The smooth muscle constructs a stomach, a bowel and a blood vessel. The skeletal muscle accounts for the greater part of muscle and moves body with bones. Movements of cardiac muscle and smooth muscle are expressed as twitch movements. The skeletal muscle movements are expressed as gross movements mostly. In this paper, we aim at measuring the gross movements. In order to measure the gross movements and evaluate them quantitatively, it is necessary to track human motions.

3 Motion Tracking System

3.1 Required Functions

In order to track bed-ridden person's motions, three functions described below are required.

Unrestraint Sensing: Attaching sensors to the body produces mental and physical burdens for persons. In order to measure unaffected body motions, sensors need to be attached not to the body but to an environmental side as a bed. In this paper, a pressure sensors distribution bed is introduced to measure body motions unrestraintly.

Personal Fitting: The shapes of the bodies are various for each person. If the shape of the body changes, the pressure distribution, which is measured by the bed, also changes. In this paper, in order to cope with a personal variety of the shape of the body, a surface model, which fits the bed-ridden person, is imported to the tracking system. By using the imported surface model, the personal fitting function is thought to be realized.

Vertical Motion Tracking: In order to track human vertical motions such as raising the upper body, a skeleton and a surface integration model is introduced. By using this integration model, the shape of the surface can be changed according to the joint angles of the skeleton. A pressure distribution image is constructed based on this integration model and is compared with the measured pressure distribution image. Considering with potential energy, momentum and a difference between the model based and measured images, the vertical motion tracking function is realized.

3.2 System Overview

A human motion tracking system that satisfies with required functions described in section 3.1 is realized. This system consists of a pressure sensors distribution bed (Figure 1 (a)), pressure sensors control box (Figure 1 (b)), a pressure distribution measuring computer (Figure 1 (c)) and a human motion tracking computer (Figure 1 (d)).

The pressure distribution measuring computer measures the pressure distributions of the bed-ridden person and transmits them to the human motion tracking computer. This computer tracks human motion by using following algorithm.

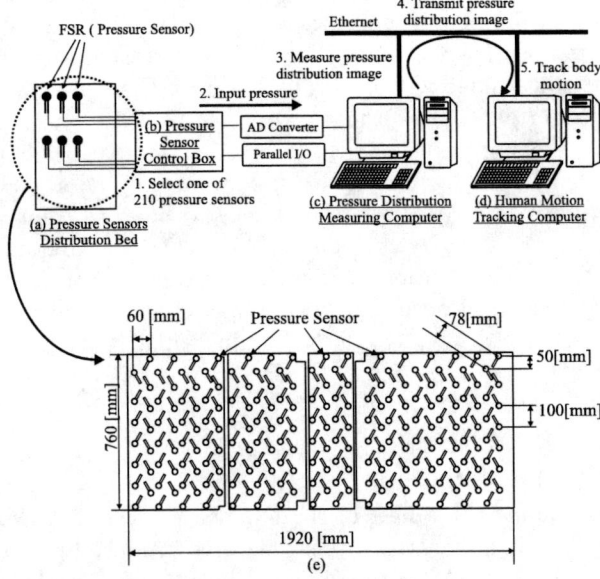

Figure 1: Hardware configuration

1. Importing a skeleton model and a surface model

2. Manual adjustment of the skeleton model and the surface model

3. Construction of an initial full body model by integrating the skeleton model and the surface model

4. Measuring a pressure distribution as a motion tracking target at the frame t

5. Construction of a pressure distribution image by using a full body model

6. Calculation of joint and translation parameters considering a difference between measured and model based pressure distribution, gravity and momentum

7. Updating joint and translation parameters of the full body model and re-calculating a model based pressure distribution image

8. If difference between measured and model based pressure distribution converges, then decide parameters at frame t and return to a pressure distribution image measuring at frame $t+1$. Otherwise return to the joints and translation parameters calculation for searching better parameters at frame t.

Details of this system are described in the following sections.

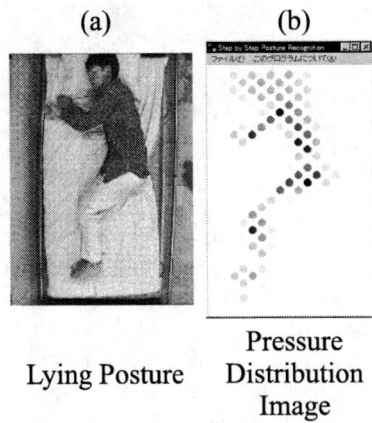

Lying Posture   Pressure Distribution Image

Figure 2: Pressure distribution image

3.3 Pressure Sensors Distribution Bed

The pressure sensors distribution bed has 210 arrayed pressure sensors. Figure 1 (e) shows the bed. Force Sensing Register (FSR) is used as the pressure sensor. The FSR is a thin film sensor, which is made from piezoresistive polymer. The resister value of the FSR is reduced in proportion to applied force. The size of the pressure sensors distribution bed is 1920 × 760 × 17 [mm]. A distance between pressure sensors is 78 [mm]. About 50[mm] thickness futon mattress is spread over this sensor bed.

The pressure sensors control box can select one of the pressure sensors and read the selected pressure sensor's value. The pressure sensors control box scans pressure sensors one by one by using multiplexers, reads one of pressure sensors' value and transmits this sensor's value to the pressure distribution measuring computer. This computer measures pressure distribution as a pressure distribution image. Sampling frequency of the image is about 10 [Hz]. Its resolution is 12 bit. Figure 2(a) shows the lying person's picture and Figure 2(b) shows the pressure distribution image. The dark color indicates a high pressure point.

3.4 Full Body Model

A full body model is constructed by integrating a skeleton model and a surface model. The following sections explain the skeleton and surface model and an integration method of them.

3.4.1　Skeleton Model

BVH files are used as skeleton models. BVH file can be divided into a hierarchy part and a motion part. The hierarchy part describes a structure of joints and links. The motion part describes time series of joints and translation parameters. The system utilizes the hierarchy part of BVH file to construct the skeleton model.

Merits of using BVH files are: 1) Since much software can import and export BVH files, by using these software it is easy to modify the structure of the skeleton. 2) Importing and exporting BVH files is very easy because BVH file is described as ASCII format. 3) Tracking results can be analyzed anywhere if these results are exported as the BVH format.

As one of the examples of BVH files, Figure 3 shows the structure of the skeleton and joints' name, which is included in Biovision motion collections. The root of this skeleton is the hip. The number of joints is 18. Each joint has three degree of freedom. The hip has six degree of freedom. The total number of degree of freedom is 57.

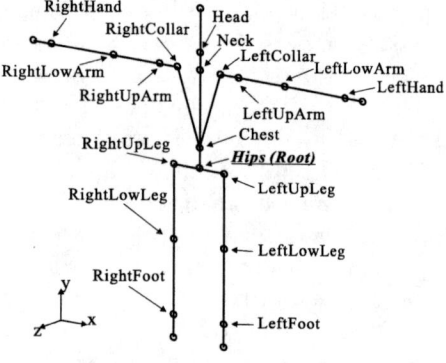

Figure 3: BVH File from Biovision Motion Collection

3.4.2　Surface Model

In order to realize the personal fitting functions, a surface model is adopted. Although there are many formats, which describe surface models, in this paper Object file of Wavefront's Advanced Visualizer is adopted. Objects files define the geometry and other properties for objects. Object files can be in ASCII format (*.obj) or binary format (*.mod). Since it is easy to handle the ASCII format files, the obj file is adopted.

Merits of using obj files as the surface models are: 1) Since much software can import and export to obj files, by using these software it is easy to modify the

structure of the surface. 2) It is possible to correspond to the different shape of the body only by changing obj files. 3) There is high possibility of utilizing formats that are different from obj files because there are many converters to obj format.

The obj file is not the solid model, but is the hollow model. The reason why the hollow model is adopted is described below. There is an elastic body as a futon mattress between the human body and the pressure sensors distribution bed. Reaction force is thought to be produced according to a displacement. The relationship between the human and the mattress is given by:

$$mg + F = f(x) \qquad (1)$$

where m is the mass of the human, x is the displacement of the mattress, F is the force produced by the muscle and $f(x)$ is the force produced by the displacement of the mattress. If the force applying to the pressure sensors is defined as P, the relationship between the mattress and the pressure sensor is given by:

$$P = f(x) \qquad (2)$$

The equation eliminating $f(x)$ from equation (1) and equation (2) is given by:

$$mg + F = P \qquad (3)$$

Equation (3) expresses that the force applying to the pressure sensors can be calculated, if we know the mass of the body and the force produced by the muscle. However, it is very complicated to construct the model including the mass and the force. On the other hand, from the equation (2), the force applying to the pressure sensor is expressed by the function of the distance between the pressure sensor and the human body. The force applying to the pressure sensor can be calculated by the knowledge of the distance of the pressure sensor and the human body. Therefore, in order to construct the model based pressure distribution image, it is much easier to use the hollow model than the solid model.

3.4.3　Integration of Skeleton and Surface

After importing the skeleton model and the surface model, these models are integrated into the initial full body model. Before integration, the system scales the skeleton model and the surface model automatically to the bed-ridden person's height, which is previously inputted. It is necessary to adjust the skeleton model

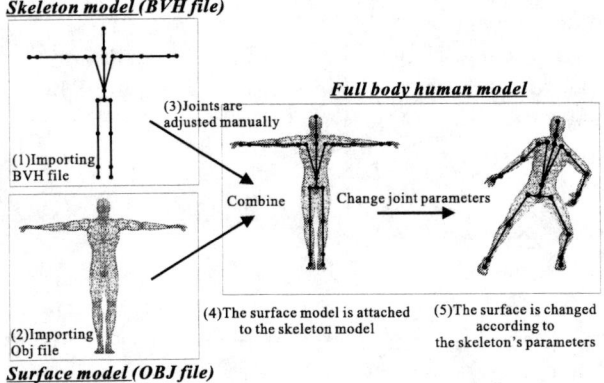

Figure 4: Integration of Skeleton and Surface

and the surface model manually, because there is no relationship between the BVH file and the obj file. The skeleton model is moved and deformed in order that the skeleton's joints coincide with the surface's joints.

The surface model is assigned automatically to the skeleton model after manual adjustment of the skeleton model. The assignment algorithm is described below. First, one of the vertexes, which construct the surface model, is selected. Secondly, a distance between the selected vertex and all the links of the skeleton is calculated. Thirdly, the vertex is assigned to the link whose distance is the shortest. These assign procedures are operated for all the vertex. After this assignment, the surface model can change according to the skeleton's joint and translation parameters. Figure 4 shows this assignment procedure.

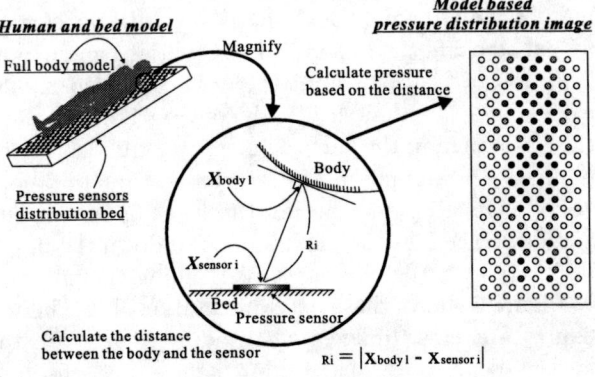

Figure 5: Model based pressure distribution

3.5 Model Based Pressure Distribution

It can be thought that pressures depend on the distance between the pressure sensors and the surface as described in section 3.4.2. We assume that the shortest distance between the pressure sensor and the surface influences the pressure of this sensor.

We consider the relationship between the distance and the pressure. If the force applying to the mattress is small, the displacement of the mattress is in proportion to the force. However if the force increases, the rate of the displacement becomes small. Therefore, the force function of the mattress is non-linear whose characteristics are: 1) The force is 0, if the surface is far from the sensor. 2) If the surface is close to the sensor, the pressure is very high. One of the simplest functions, which contain these characteristics, is the following function.

$$w_i = \frac{P_a}{(r_i + P_b)^2} \qquad (4)$$

where w_i is the pressure of the ith sensor, r_i is the shortest distance between the ith sensor and the surface, P_a and P_b are coefficients. A set of $w_i(1 \le i \le N_{sensor})$ is the pressure distribution image where N_{sensor} is the number of the pressure sensors. This function is applied to construct the model based pressure distribution image.

P_a and P_b are decided by comparing with the measured pressure distribution images and are fixed irrespective of the different shape of the body. Assuming that meshes of the surface model are very small, r_i is calculated by:

$$r_i = \min_l \{||\boldsymbol{v}_l - \boldsymbol{s}_i||\} \qquad (5)$$

where $\boldsymbol{v}_l$ is the lth position vector of the vertex of the surface, $\boldsymbol{s}_i$ is the position vector of the ith sensor. Figure 5 shows the model based pressure distribution image construction process.

3.6 Tracking Method

In order to track the human motion, joint and translation parameters are determined to minimize the difference between the measured pressure distribution image and the model based pressure distribution image. This difference is given by:

$$E_p = \sum_{i=1}^{N_{sensor}} (w_i - m_i)^2 \qquad (6)$$

103

where m_i is the measured pressure of ith sensor. A set of $m_i(1 \leq i \leq N_{sensor})$ is the measured pressure distribution image.

The body is influenced by the gravity. If no force is applied to the body, the body falls to the bed. The potential energy of the gravity is given by:

$$E_g = \sum_{j=1}^{N_{node}} gz_j \qquad (7)$$

where z_j is the height of the skeleton's j^{th} node from the bed, N_{node} is the number of the nodes of the skeleton and g is the gravity acceleration. As an evaluation function for tracking the human motion, the sum of the square of the difference and the potential energy is adopted. The evaluation function E_{all} is given by:

$$E_{all} = \epsilon_p E_p + \epsilon_g E_g \qquad (8)$$

where ϵ_p is weight for the difference and ϵ_g is weight for the potential energy. The skeleton's joint and translation parameters are determined to minimize this evaluation function. We assume that the parameters are searched sequentially. This process is given by:

$$\frac{dx_k}{dt} = -\frac{\partial E_{all}}{\partial x_k} \qquad (9)$$

where x_k is the kth skeleton's joint and translation parameter. Considering with equation (6) and equation (7), the equation (9) can be rewrote as:

$$\frac{dx_k}{dt} = -\epsilon_k \{ 2\epsilon_p \sum_i (w_i - m_i) \frac{\partial w_i}{\partial x_k} + \epsilon_g \sum_j g \frac{\partial z_j}{\partial x_k} \} \quad (10)$$

where ϵ_k is added as weight for the differential of x_k. Then the differential of x_k is transferred to discrete-time expression as $\triangle x_k(t, n)$. Where t is the frame number of the measured pressure distribution image and n is the calculation time for convergence. In order to accelerate convergence, momentum term is added to $\triangle x_k(t, n)$. The $\triangle x_k(t, n)$ can be rewrote as:

$$\triangle x'_k(t, n) = \triangle x_k(t, n) + \alpha \triangle x'_k(t, n-1) \qquad (11)$$

where $\alpha(0 < \alpha < 1)$ is weight for momentum term. Then the skeleton's parameters are given by:

$$x_k(t, n) = x_k(t, n-1) + \triangle x'_k(t, n) \qquad (12)$$

These processes are calculated until the evaluation function converges.

At the first time of the convergence calculation ($n = 1$), we think that $\triangle x'_k(t+1, 1)$ is influenced by the parameters at frame t and frame $t-1$. Therefore, $\triangle x'_k(t+1, 1)$ is given by:

$$\triangle x'_k(t+1, 1) =$$
$$\triangle x_k(t+1, 1) + \beta\{x_k(t) - x_k(t-1)\} \quad (13)$$

where $x_k(t)$ is the final state of kth parameter at frame t and $\beta(0 < \beta < 1)$ is weight for the influence of the difference between k^{th} parameters at frame t and frame $t-1$. The initial state of skeleton's joints and translation parameters at frame $t+1$ is the final state at frame t.

$$x_k(t+1, 0) = x_k(t) \qquad (14)$$

At the initial state of frame 1, skeleton's joints and translation parameters ($x_k(1, 1)$) are adjusted to match the lying person's posture manually at the moment.

4 Experiment

A subject is a male whose height is 176 cm and weight is 61 kg. The minimum number of the backbone joints is one and its degree of freedom is three for expressing the upper body upward and downward motion or twist motion. Therefore, we use the BVH file that is included in the Biovision motion collection as mentioned Figure 3, although the degree of freedom of this model's backbone is fewer than the degree of freedom of the human's backbone. We create the surface model by using the Poser3 (MetaCreation) and export it to the obj file. The number of vertexes of this model is 2404 and the number of polygons is 4124. By using this skeleton and the surface integration model, we experimented a human motion tracking. Target motions are: 1) Opening and closing the legs. 2) Moving up and down the legs. 3) Moving up and down the upper body. 4) Twisting the body. 5) Bending the knees.

Figure 6 shows these experimental results. The top figures are video images, the middle figures are the measured pressure distribution images and the bottom figures are tracking results of the full body model. These experimental results show that human motions mentioned above can be tracked correctly by using the proposed algorithm. However the model's lower arms and hands positions does not coincide with the video images slightly. That is because the lower arms and hands do not appear on the measured pressure distribution images.

5   Conclusion

In this paper, we developed unrestraint human motion tracking system by using the skeleton and the surface integration model and the pressure sensors distribution bed. By using the pressure sensors distribution bed, the bed-ridden person's unaffected body movement can be measured. Thanks to the integration of the surface model and the skeleton model, the pressure distribution image can be calculated. Considering with the difference between the model based pressure distribution image and the measured pressure distribution image, potential energy and the momentum of the body movement, motions such as upward and downward motion and twist motions of which tracking has not been realized can be tracked correctly. This system is thought to be used for analyzing the body movement, a health monitoring and evaluation of rehabilitation progress and so on. Automatic initial position adjustment, different shape of body experiments and clarifying the limitation of the tracking are future works.

Acknowledgments

The authors would like to thank Mr. Yoshimi for developing the pressure sensors distribution bed. This research is supported by JSPS Grant-in-Aid for Scientific Research No.12-08887 and No.11555069.

References

[1] Alihanka J, Vaahtoranta K, Saarikivi I, "A New Long-term Monitoring of Ballistocardiogram, Heart Rate, and Respiration," AM J Physiol., Vol.240, pp.384-392, 1981.

[2] Toshiyo Tamura, Jiangxin Zhou, Hiroshi Mizukami and Tatsuo Togawa, "A System for Monitoring Temperature Distribution in Ded and Its Application to The Assessment of Body Movement," Physiol. Meas., Vol.14, pp.33-41, 1993.

[3] Yoshifumi Nishida, Masashi Takeda, Taketoshi Mori, Hiroshi Mizoguchi and Tomomasa Sato, "Monitoring Patient Respiration and Posture Using Human Symbiosis System, in Proc. of International Conference on Intelligent Robots and Systems, Vol.2, pp.632-639, 1997.

[4] Tatsuya Harada, Taketoshi Mori,Yoshifumi Nishida, Tomohisa Yoshimi and Tomomasa Sato, "Body Parts Positions and Posture Estimation System Based on Pressure Distribution Image," Proc. of 1999 IEEE International Conference on Robotics and Automation, Vol.2, pp.968-975, 1999.

[5] Tatsuya, Harada, Akihiko Saito, Tomomasa Sato and Taketoshi Mori, "Infant Behavior Recognition System Based on Pressure Distribution Image," Proceedings of the 2000 IEEE International Conference on Robotics and Automation, Vol.4, pp.4083-4089, 2000.

[6] Alan K. Nakahara, David L. Jaffe, and Eric E. Sabelman, "Development of a Second Generation Wearable Accelerometric Motion Analysis System," Proc. of 2nd National Rehabilitation Research and Development Service Meeting, 2000.

[7] Joseph O'Rourke and Norman I. Badler, "Model-Based Image Analysis of Human Motion Using Constraint Propagation," IEEE Trans. Pattern Analysis and Machine Intelligence, Vol. PAMI-2, No.6, pp.522-536, November 1980.

[8] J. Rehg and T. Kanade, "Visual Tracking of High DOF Articulated Structures: An Application to Human Hand Tracking," Proc. of Third European Conf. on Computer Vision, Vol. II, pp. 35-46, 1994.

[9] Christopher R. Wren, Brian P. Clarkson, and Alex P. Pentland, "Understanding Purposeful Human Motion," Proceedings of the Fourth IEEE International Conference on Automatic Face and Gesture Recognition, Grenoble, France, March 26-30, 2000.

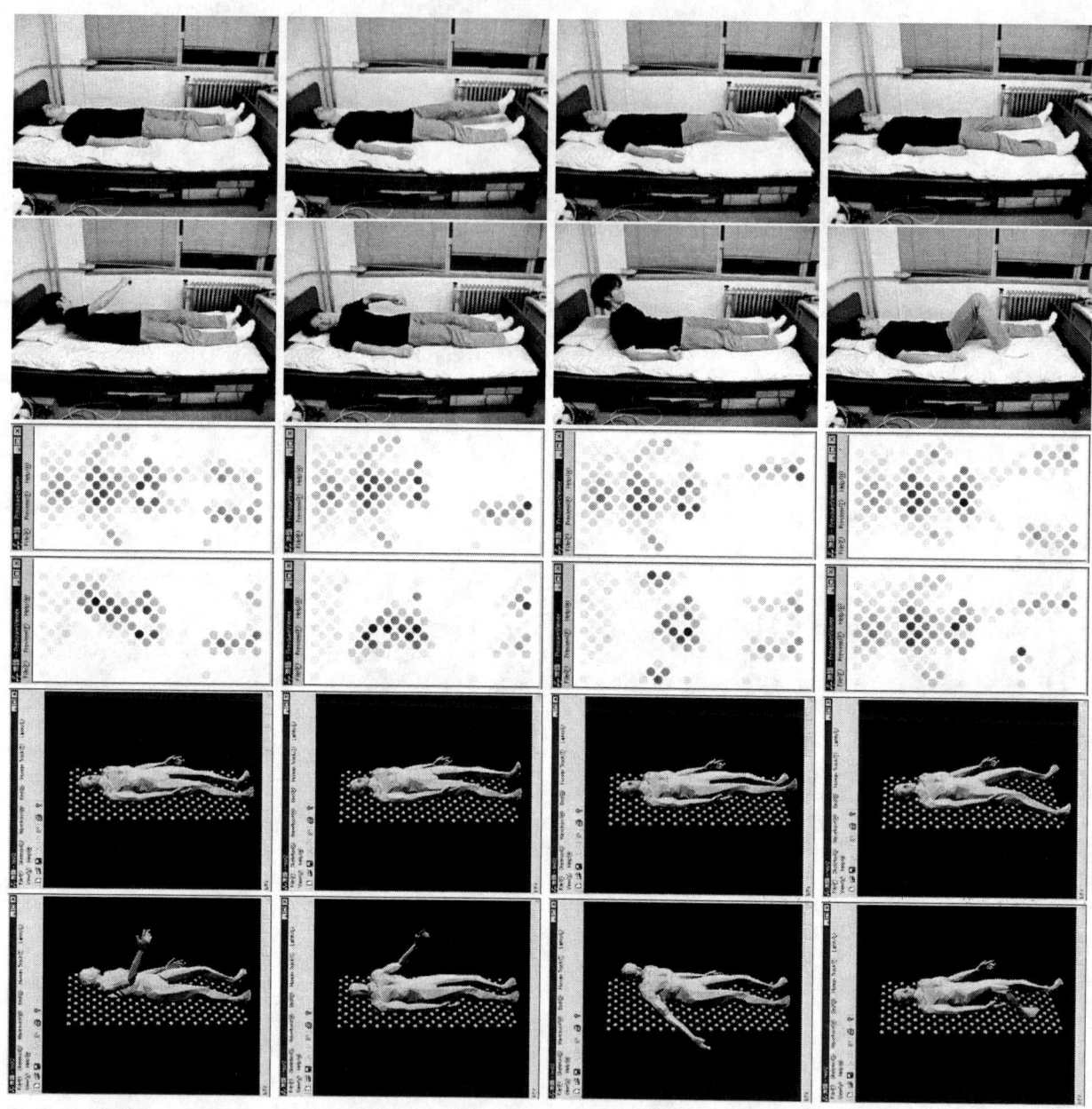

Figure 6: Experimental results. Top figures: Video images of lying person. Middle figures: Pressure distribution images. Bottom figures: Tracking motions of the full body model.

Activity monitoring and summarization for an intelligent meeting room

Ivana Mikić, Kohsia Huang, Mohan Trivedi
Computer Vision and Robotics Research Laboratory
Department of Electrical and Computer Engineering
University of California, San Diego

Abstract

Intelligent meeting rooms should support efficient and effective interactions among its occupants. In this paper, we present our efforts toward building intelligent environments using multimodal sensor network of static cameras, active (pan/tilt/zoom) cameras and microphone arrays. Active cameras are used to capture details associated with interesting events. The goal is not only to make the system that supports multiperson interactions in the environment in real time, but also to have the system remember the past, enabling review of past events in an intuitive and efficient manner. In this paper, we present the system specifications and major components, integration framework, active network control procedures and experimental studies involving multiperson interactions in an intelligent meeting room environment.

1. Introduction

Intelligent environments are a very attractive domain of investigation due to both the exciting research challenges and the importance and breadth of possible applications. It is strongly influencing recent research in computer vision [1]. Realization of such spaces requires innovations not only in the computer vision [2, 3, 4, 5], but also in audio-speech processing and analysis [6, 7] and in the multimodal interactive systems area [8, 9, 10].

In this paper, we describe the system that handles multiperson interactions in an intelligent meeting room – Figure 1. It is being developed and evaluated in a multipurpose testbed called AVIARY (Audio-Video Interactive Appliances, Rooms and sYstems) that is equipped with four static and four active (pan/tilt/zoom) rectilinear cameras and two microphones.

2. Intelligent meeting room (IMR)

We consider IMRs to be spaces which support efficient and effective interactions among their human occupants. They can all be occupying the same physical space or they can be distributed at multiple/remote sites. The infrastructure which can be utilized for such intelligent rooms include a suite of multimodal sensory systems, displays, pointers, recording devices and appropriate computing and communications systems. The necessary "intelligence" of the system provides adaptability of the environment to the dynamic activities of the occupants in the most unobtrusive and natural manner.

Figure 1. An intelligent meeting room

The types of interactions in an intelligent environment impose requirements on the system that supports them. In an intelligent meeting room we identify three types of interactions:

- between active participants – people present in the room
- between the system and the remote participants
- between the system and the "future" participants

The first category of interactions defines the interesting events that the system should be able to recognize and capture. The active participants do not obtain any information from the system but cooperate with it, for example by speaking upon entering the room to facilitate accurate person identification.

Other two types of interactions are between the system and people that are not present in the room. Those people are the real users of the system. For the benefit of the remote participant, the video from active cameras that capture important details such as a face of the presenter or a view of the whiteboard should be captured and transmitted. Information on identities of active participants, snapshots of their faces and other information can be made available. The "future" participant, the person reviewing the meeting that happened in the past, requires a tool that graphically summarizes past events to easily grasp the spatiotemporal relationships between events and people that participated in them. Also an interface for interactive browsing and review of the meeting is desirable. It would provide easy access to stored information about the meeting such as identities and snapshots of participants and video from active cameras associated with specific events.

Interactions between active participants in a meeting room define interesting activities that the system should be able to recognize and capture. We identified three: a person located in front of the whiteboard, a lead presenter speaking and other participants speaking. A lead presenter is the person currently in front of the whiteboard. First activity should draw attention from one active camera that captures a view of the whiteboard. Other two activities draw attention from an active camera with the best view of the face for capturing the video of the face of the current speaker.

To recognize these activities, the system has to be aware of the identities of people, their locations, identity of the current speaker and the configuration of the room. Basic components of the system that enable described functionality are:

- 3D tracking of centroids using static cameras with highly overlapping fields of view
- Person identification (face recognition, voice recognition and integration of the two modalities)
- Event recognition for directing the attention of active cameras
- Best view camera selection for taking face snapshots and for focusing on the face of a current speaker
- Active camera control
- Graphical summarization/user interface component

Details associated with the overall architecture and specific components for IMR are in the next section.

3. The IMR components and system architecture

Integration of audio and video information is performed at two levels. First the results of face and voice recognition are integrated to achieve robust person identification. At a higher level, results of 3D tracking, voice recognition, person identification (which is itself achieved using multimodal information) and knowledge of the structure of the environment are used to recognize interesting events.

When a person enters the room, the system takes the snapshot of their face and sample of their speech to perform person identification using face and voice recognition [11, 12].

The system block diagram is shown in Figure 2. As mentioned before, it currently takes inputs from four static cameras with highly overlapping fields of view, four active cameras and two microphones. All of the eight cameras are calibrated with respect to the same world coordinate system using Tsai's algorithm [13].

Two PC computers are used. One performs 3D tracking of blob (people and objects) centroids based on input from four static cameras. Centroid, velocity and bounding cylinder information is sent to the other PC which handles all other system functions. For new people in the environment, the camera with the best view of the face is chosen and moved to take the snapshot of the face. The person is also required to speak at that time and the system combines face and voice recognition results for robust identification. Identity of the current speaker is constantly monitored and used to recognize interesting events together with 3D locations of people and objects and known structure of the environment. When such events are detected, the attention of active cameras is directed toward them.

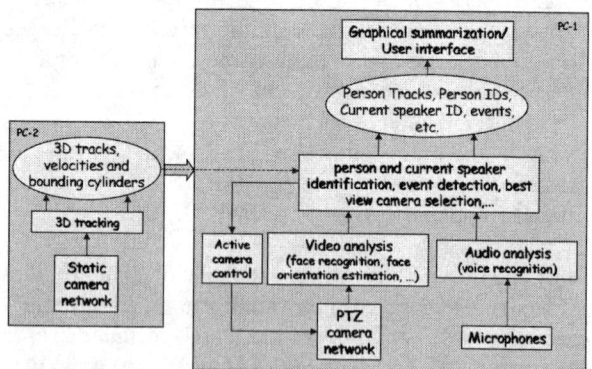

Figure 2. Block diagram of the system

3D centroid tracking. Segmentation results (object centroids and bounding boxes) from each of the four static cameras are used to track centroids of objects in the room and their bounding cylinders in 3D. Details of the tracking algorithm are given in [14]. The tracker is capable of tracking multiple objects simultaneously. It maintains a list of Kalman filters, one for each object in the scene. The tracker calculates updated and predicted positions for each object in real time. Availability of up-to-date predictions allows feedback to the segmentation algorithm, which can increase its sensitivity in the areas where the objects are expected to appear. Figure 3 shows a typical input from four cameras with object centroids and bounding boxes calculated by the segmentation algorithm. Projections of the tracks back onto image planes and also projections onto the floor plane are shown.

Figure 3. 3D tracking – smaller crosshairs (barely visible) and green bounding boxes are segmentation results used by the tracker to compute 3D tracks and bounding cylinders. Larger crosshairs are projections of tracks back onto the image planes. Bottom: projections of the track onto the floor plane.

Person identification and current speaker recognition. Eigenface recognition algorithm [15] is currently utilized in the face recognition module. Human face is extracted from the snapshot image of camera network by skin color detection [16]. Face images of known people at certain facing angles are stored into the training face database. The face image is compared to the training faces in terms of distances in the eigenface space. The test face is then classified as a known person if the minimum distance to the corresponding training face is smaller then the recognition bound. For voice recognition, we use a text independent speaker identification module from the IBM ViaVoice SDK. When there is speech activity, clips of up to 5 seconds in length are recorded and sent to ViaVoice for recognition

The results of face and speaker recognition modules are fused together for robust person identification. Since ViaVoice does not provide access to confidence measures

of recognition results, we are not able to make optimal decisions. Therefore, we perform the following fusing scheme. Each module gives output only if there is reasonable confidence associated with it. If only one module outputs a valid result, then it is taken as the final decision. If both modules output valid but different results, the output from face recognition is accepted if its confidence is above predetermined high value, otherwise the output from speaker recognition is accepted.

Event recognition for directing the attention of active cameras. This module constantly monitors for events described in the section 2. When a new track is detected in the room, it is classified as person or object depending on the dimensions of the bounding cylinder. This classification is used to permanently label each track. If classified as object, the camera closest to it takes the snapshot. If classified as person the camera with the best view of the face needs to be selected. The snapshot is then taken and person identification is performed. Each person track is labeled with person's name. Events are associated with tracks labeled as people (person located in front of a whiteboard, person in front of the whiteboard speaking and person located elsewhere speaking) and are easily detected using track locations and identity of the current speaker.

Best view camera selection. The best view camera for capturing the face is the one for which the angle between the direction the person is facing and the direction connecting the person and the camera is the smallest (Figure 4). Center of the face is taken to be 20cm from the top of the head (which is given by the height of the bounding cylinder).

There are three different situations where the best view camera selection is performed. First is taking snapshot of the face of the person that just entered the room. Second, if the person in front of the whiteboard is speaking a camera needs to focus on their face. The third situation is when the person not in front of the whiteboard speaks. In these three situations, we use different assumptions in estimating the direction the person is facing.

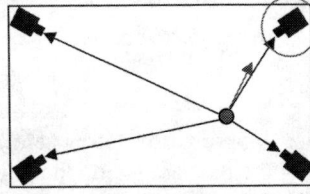

Figure 4. Best view camera is chosen to be the one the person is facing the most (maximum inner product between the direction the object is facing and direction toward a camera)

When a person walks into the room, we assume that they are facing the direction in which they are walking. If a person is in front of a whiteboard (location of which is known), one camera focuses on the whiteboard (Figure 5). If the person starts speaking, a best view camera needs to be chosen from the remaining cameras to focus on that person's face. Since the zoomed-in whiteboard image contains person's head, we use that image to estimate the direction the person is facing. Due to the hairline, the ellipse fitted to the skin pixels changes orientation as person turns from far left to far right (Figure 6). We use skin detection algorithm described in [16]. If skin pixels are regarded as samples from a 2D Gaussian distribution, the eigenvector corresponding to the larger eigenvalue of the 2×2 covariance matrix describes the orientation of the ellipse. A lookup table based on a set of training examples (Figure 7) is used to determine the approximate angle between the direction the person is facing and the direction connecting the person and the camera that the whiteboard image was taken with. These angles are not very accurate, but we have found that this algorithm works quite reliably for purposes of best view camera selection.

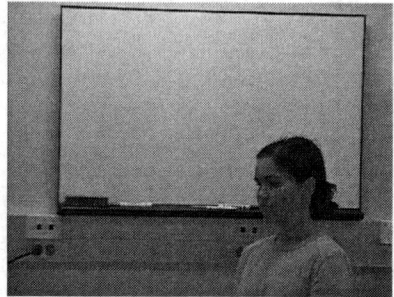

Figure 5. Person close to the whiteboard draws attention from one active camera

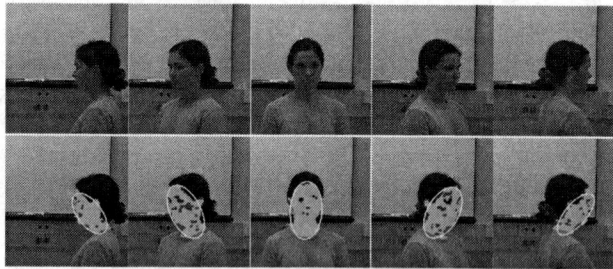

Figure 6. Face orientation estimation for best view camera selection

In the third case, where person elsewhere in the room is speaking, we assume they are facing the person in front of the whiteboard if one is present there. Otherwise, we assume they are facing the opposite side of the room. The first image obtained with the chosen camera is processed using the algorithm described in the previous paragraph and the camera selection is modified if necessary.

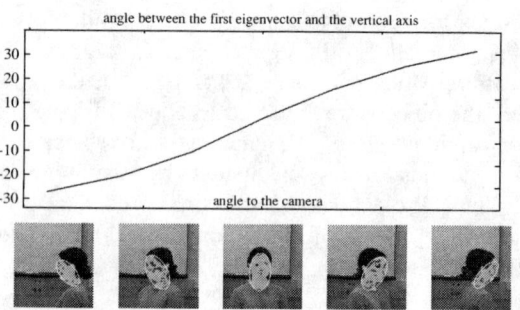

Figure 7. Lookup table for face orientation estimation computed by averaging across training examples

Active camera control. Pan and tilt angles needed to bring the point at a known location to the center of the image can be easily computed using the calibrated camera parameters. However, the zoom center usually does not coincide with the image center. Therefore, the pan and tilt angles needed to direct the camera toward the desired location have to be corrected by the pan and tilt angles between the center of the image and the zoom center. Otherwise, for large magnifications, the object of interest may completely disappear from view.

A lookup table (Figure 8) is used to select a zoom needed to properly magnify the object of interest (person's face or a whiteboard). Zoom magnification is calibrated using the model and the algorithm described in [17]:

$$x' = M(n)[x - C_x] + C_x$$
$$y' = M(n)[y - C_y] + C_y \qquad (1)$$

where n is the current zoom value, $M(n)$ is the magnification, C_x and C_y are the coordinates of the center of expansion or the zoom center, x and y are the coordinates of a point at zero zoom and x' and y' are coordinates of the same point at zoom n.

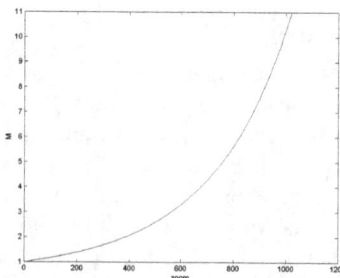

Figure 8. Magnification for different zoom values for on of the active cameras

Magnifications are computed for a subset of possible zoom values defined by a chosen zoom step. Magnifications for other zoom values are interpolated from the computed ones. The magnifications are obtained using a slightly modified version of [17]. Two images taken with two different zoom values are compared by shrinking the one taken with the larger zoom using the Equation 1. The value of magnification (will be smaller than 1) that achieves best match between the two images is taken to be the inverse of the magnification between the two images. The algorithm described in [17] was written for outdoor cameras where objects present in the scene are more distant from the camera than in the indoor environments. Therefore, instead of comparing images at different zooms to the one taken at zero zoom as done in [17], we always compare two images that are one zoom step apart. The absolute magnification for a certain zoom value with respect to zero zoom is computed by multiplying magnifications for smaller zoom steps (Figure 8). However, we could not reliably determine the location of the zoom center using this algorithm. Instead, we determine its coordinates manually by overlaying a crosshair over the view from the camera and zooming in and out until we find a point that does not move under the crosshair during zooming.

Graphical summarization/user interface. The history is summarized graphically for easy review and browsing of information the system collected about the environment. The 3D graphic representation shows the room floor plan and the third axis represents time (**Figure 9**). Tracks are color-coded and represented by one shape (i.e. sphere) when the person is not speaking and by a different one (i.e. cube) when the person is speaking. The floorplan shows important regions like whiteboard and doors. This graphical representation effectively summarizes events that system can detect and trajectories and identities of people involved. It also serves as a user interface. By clicking on a colored shape, the user is shown the face snapshot and the name of the person associated with the track and the video associated with the event the shape corresponds to can be replayed.

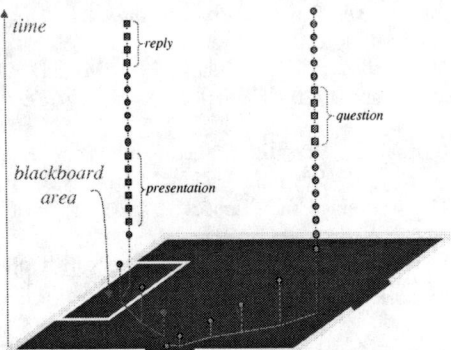

Figure 9. Graphical summarization of the events in the environment. A presenter (red) and another participant (green) were present.

4. System performance

The described system is operating quite reliably. In [14], we have described the experiments on the accuracy of

centroid tracking and have reported good results with maximum errors around 200mm. We currently have only five people in the face and speaker databases, so the person identification accuracy based on both modalities is practically 100%. Also, recognition of the current speaker performs with nearly perfect accuracy if silence in a speech clip is less then 20% and clip is longer than 3 seconds. The results are very good for clips with low silence percentage even for shorter clips, but gets erroneous when silence is more than 50% of the clip. However, there is a delay of 1-5 seconds between the beginning of speech and the recognition of the speaker, which causes a delay in recognizing activities that are concerned with the identity of the current speaker.

If the person faces the direction they are walking, the camera selection for acquisition of face snapshots also works with perfect accuracy. It would, of course, fail if person turned their head while walking. The camera selection for focusing on the face of the person that is talking in front of the whiteboard succeeds around 85% of the time. In the case of the person talking elsewhere in the room, our assumption that they are facing the person in front of the whiteboard or the opposite side of the room is almost always true. This is due to the room setup – there is one large desk in the middle of the room and people sit around it – therefore almost always facing the opposite side of the room, unless they are talking to the presenter

We can store all information needed to access appropriate parts of the video that correspond to the events the user selects from the interface. From the interface, the user can view identities and face snapshots of people associated with different tracks by clicking on the corresponding colored shape. For remote viewing, the videos from active cameras that capture interesting events can be transmitted together with the other information needed to constantly update the summarization graph. See Figure 10 for the illustration of the system operation.

5. Concluding remarks

We have presented our investigations toward building the multimodal intelligent environments that provide awareness of people and events at several resolution levels: from a graphical summarization of the past and ongoing events to the active camera focus on interesting events and people involved in them.

Next step in this investigation would be more detailed and sophisticated audio and video analysis that would use high-resolution information collected by the active camera and microphone network. This would include posture estimation, gesture recognition, speech recognition, lip-reading, etc.

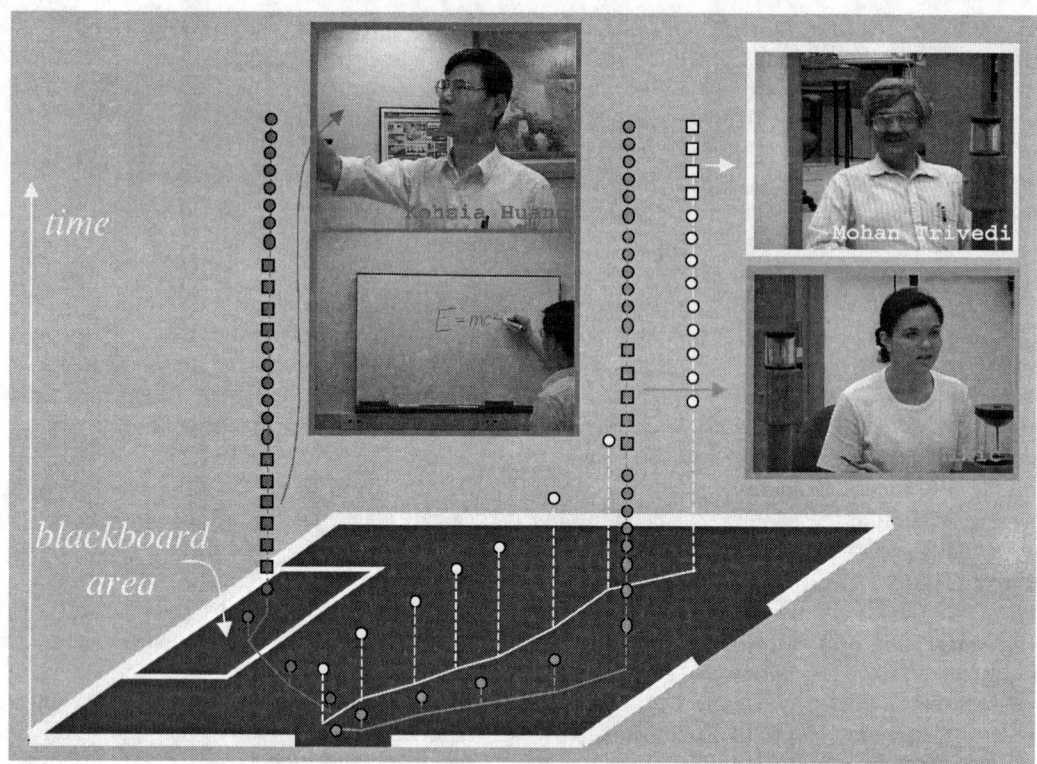

Figure 10. Illustration of the system operation. Interesting activities attract attention from active cameras. That video can be transmitted to remote viewers or stored for later reviewing. Every "object" in this graphical summarization is associated with information needed to access the appropriate portion of video, face snapshots and identity information

References

[1]   A. Pentland, "Looking at People: Sensing for Ubiquitous and Wearable Computing", *IEEE. Trans. PAMI*, 22(1), Jan 2000, pp. 107-119

[2]   D. Gavrila, "The Visual Analysis of Human Movement: A Survey", *Computer Vision and Image Understanding*, 73(1), Jan 1999, pp. 82-98

[3]   V. Pavlović, R. Sharma, T. Huang, "Visual Interpretation of Hand Gestures for Human-Computer Interaction: A Review", *IEEE Trans. PAMI*, 19(7), July 1997, pp. 677-695

[4]   R. Cipolla, A. Pentland (editors), *Computer Vision for Human-Machine Interaction*, Cambridge University Press, Cambridge, UK, 1998

[5]   R. Chellappa, C. Wilson, S. Sirohev, "Human and Machine Recognition of Faces: A Survey", *Proc. IEEE*, 83(5), pp. 705-740, 1995.

[6]   L. Rabiner, B. Juang, "Fundamentals of Speech Recognition", Englewood Cliffs, NJ: Prentice-Hall 1993.

[7]   M. Brandstein, J. Adcock, H. Silverman, "A closed-form location estimation for use with room environment microphone arrays", *IEEE Trans. Speech and Audio Processing*, 5(1), Jan. 1997, pp. 45-50

[8]   R. Sharma, V. Pavlović, T. Huang, "Toward Multimodal Human-Computer Interface", *Proc. IEEE*, 86(1), May 1998, pp. 853-869

[9]   M. Trivedi, B. Rao, K. Ng, "Camera Networks and Microphone Arrays for Video Conferencing Applications", *Proc. Multimedia Systems Conf.*, Sep 1999

[10]   C. Wang, M. Brandstein, "A hybrid real-time face tracking system", Proc. IEEE ICASSP '98, p.3737-40

[11]   M. Trivedi, I. Mikic, S. Bhonsle, "Active Camera Networks and Semantic Event Databases for Intelligent Environments", *IEEE Workshop on Human Modeling, Analysis and Synthesis*, June 2000

[12]   M. Trivedi, K. Huang, I. Mikic, "Intelligent Environments and Active Camera Networks", *IEEE Conf. SMC 2000*

[13]   R. Tsai, "A Versatile Camera Calibration Technique for High-Accuracy 3D Machine Vision Metrology Using Off-the-Shelf TV Cameras and Lenses", *IEEE J. Robotics and Automation*, RA-3(4), 1987

[14]   I. Mikić, S. Santini, R. Jain, "Tracking Objects in 3D using Multiple Camera Views", *Proc. ACCV2000*, Jan 2000, pp. 234-239

[15]   M. Turk, A. Pentland, "Face Recognition Using Eigenfaces," *Proc. IEEE Conf. Comput. Vis and Patt. Recog.*, *Maui, HI, USA*, pp. 586-591, Jan. 1991.

[16]   J. Yang, A. Waibel, "A Real-Time Face Tracker," *Proc. WACV'96, Sarasota, FL, USA*, 1996.

[17]   Collins, Tsin, "Calibration of an Outdoor Active Camera System", CVPR '99, Fort Collins, CO, June 1999, pp. 528-534

Camera Handoff: Tracking in Multiple Uncalibrated Stationary Cameras

Omar Javed, Sohaib Khan, Zeeshan Rasheed, Mubarak Shah
Computer Vision Lab
School of Electrical Engineering and Computer Science
University of Central Florida
Orlando, FL 32816
{ojaved, khan, zrasheed, shah}@cs.ucf.edu

Abstract

Multiple cameras are needed to completely cover an environment for monitoring activity. To track people successfully in multiple perspective imagery, one needs to establish correspondence between objects captured in multiple cameras. We present a system for tracking people in multiple uncalibrated cameras. The system is able to discover spatial relationships between the camera field of views and use this information to correspond between different perspective views of the same person. We employ the novel approach of finding the limits of field of view (FOV) of a camera as visible in the other cameras. This helps us disambiguate between possible candidates of correspondence. The proposed approach is very fast compared to camera calibration based approaches.

1. Introduction

Tracking humans is of interest for a variety of applications such as surveillance, activity monitoring and gait analysis. With the limited field of view (FOV) of video cameras, it is necessary to use multiple, distributed cameras to completely monitor a site. Typically, surveillance applications have multiple video feeds presented to a human observer for analysis. However, the ability of humans to concentrate on multiple videos simultaneously is limited. Therefore, there has been an interest in developing computer vision systems that can analyze information from multiple cameras simultaneously and possibly present it in a compact symbolic fashion to the user.

To completely cover an area of interest, it is reasonable to use cameras with overlapping FOVs. Overlapping FOVs are typically used in computer vision for the purpose of extracting 3D information. However, the purpose here is not to extract depth, rather, simply to completely cover all areas of the environment. The use of overlapping FOVs, however, creates an ambiguity in monitoring people. A single person present in the region of overlap will be seen in multiple camera views. There is need to identify the multiple projections of this person as the same 3D object, and to label them consistently across cameras for security or monitoring applications.

In related work, [1] presents an approach of dealing with the handoff problem based on 3D-environment model and calibrated cameras. The 3D coordinates of the person are established using the calibration information to find the location of the person in the environment model. At the time of handoff, only the 3D *voxel-occupancy* information is compared to achieve handoff, because multiple views of the same person will map to the same voxel in 3D. In [2], only relative calibration between cameras is used, and the correspondence is established using a set of feature points in a Bayesian probability framework. The intensity features used are taken from the centerline of the upper body in each projection to reduce the difference between perspectives. Geometric features such as the height of the person are also used. The system is able to predict when a person is about the exit the current view and picks the best next view for tracking. A different approach is described in [3] that does not require calibrated cameras. The camera calibration information is recovered by observing motion trajectories in the scene. The motion trajectories in different views are randomly matched against one another and plane homographies computed for each match. The correct homography is the one that is statistically most frequent, because even though there are more incorrect homographies than the correct one, they lie in scattered orientations. Once the correct homography is established, finer alignment is achieved by global frame alignment. In a recent paper [4], the authors present a solution based on a combination of camera geometry and image features. They detect faces to find the epipolar geometry between cameras by assuming that the centroid of face boxes are corresponding points. Moreover, they use vertical features in the image to divide the view volume into non-overlapping sets, and matching view-volumes between cameras. They further use the hue

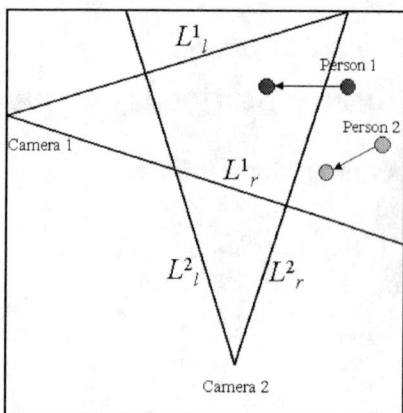

Figure 1: Person 1 and person 2 are both visible in Camera 1 initially. Person 1 then walks into the FOV of Camera 2, and needs to be identified as Person 1, for consistent labeling across views.

and saturation values of each person as features that help in establishing correspondence. Finally [5, 6] describe approaches which try to establish time correspondences between non-overlapping FOVs. The idea there is not to completely cover the area of interest, but to have motion constrained along a few paths, and to correspond objects based on time from one camera to another. Typical applications are cameras installed at intervals along a corridor [5] or on a freeway [6].

The luxury of calibrated cameras or environment models is not available in most situations. We therefore tend to prefer approaches that can discover a sufficient amount of information about the environment to solve the handoff problem. We contend that camera calibration is unnecessary and an overkill for this problem, since the only place where handoff is required is when a person enters or leaves the FOV of any camera. By building a model of only the relationship between FOV lines of various cameras can provide us sufficient information to solve the handoff problem.

In the next section we formalize the handoff problem and describe how the relationship between the FOV of different cameras can be used to solve the handoff problem. In Section 3, we describe how this relationship can be automatically discovered by observing motion of people in the environment. Finally we present results of our experiments in Section 4.

2. Edge of field of view lines

The handoff problem occurs when a person enters the FOV of a camera. At that instant we want to determine if this person is visible in the FOV of any other camera, and if so, assign the same label to this view. Consider the scenario shown in Figure 1; a room covered by two cameras with two persons walking in it. At time instant 1,

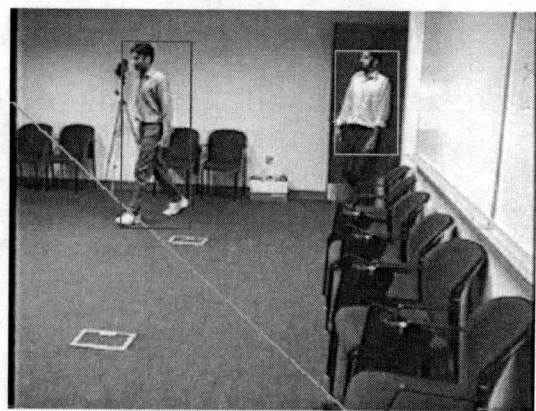

Camera 1

Camera 2

Figure 2: Example of correct handoff: There are two persons visible in Camera 1. When one of them enters the FOV of Camera 2, the left edge of FOV of Camera 2 as seen in Camera 1 (L^{21}_l) helps us disambiguate between the labels.

both persons are visible in camera 1. At time instant 2, person 1 walks into the FOV of camera 2. Since we have already assigned labels to both persons (person 1 and 2), we need to figure out at this instant which of the persons is entering the FOV of camera 2. Since we do not know any 3D information about the environment or the camera calibration matrices, we cannot determine what label to assign to the new view seen in camera 2.

Note here that we could have matched some features (e.g. color) of the two persons visible in camera 1 to the new view in camera 2 for finding the most likely match. However, when the disparity is large, both in camera location and orientation, feature matches are not reliable. After all, a person may be wearing a shirt that is different colors at front and back. The reliability of feature matching decreases with increase in disparity, and it is not uncommon to have surveillance cameras looking at an area from opposing directions. Moreover, different

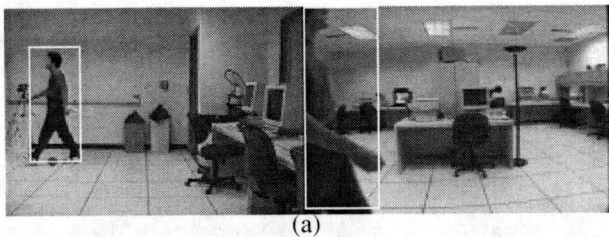
(a)

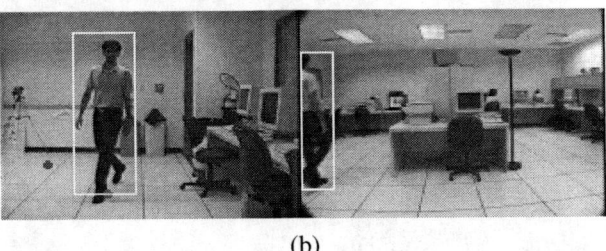
(b)

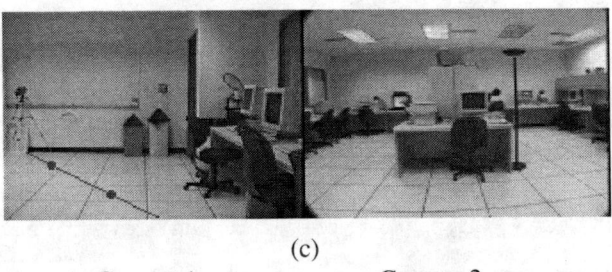
(c)

Camera 1   Camera 2

Figure 3: (a) Person entering the FOV of C^2 from left yields a point on line L^{21}_l in image taken from C^1. (b) Another such correspondence yields another point, which are joined to find the complete line L^{21}_l shown in (c).

cameras can have different intrinsic parameters as well as photometric properties (like contrast, color-balance etc.). Lighting variations also contribute to the same object being seen with different colors in different cameras.

For shallow mounted cameras each FOV's footprint can be described by two lines on the floor-plane, the left and the right limit of FOV. Let lines L^i_l and L^i_r be the left and right edge of FOV lines of the ith camera C^i on the ground plane (Figure 1). Let the projection of L^i_x ($x \in \{l, r\}$) in C^j be denoted by L^{ij}_x. Note that L^{ij}_x denotes the left or the right sides of the image in C_j. As far as the camera pair i, j is concerned, the only locations of interest in the two images for handoff are along lines L^{ij}_x and L^{ji}_x. These are up to four lines, possibly two in each camera. When a person enters the FOV of a camera, all that needs to be done is to consider the associated line in the other camera and determine which person is crossing that line. Figure 2 describes this situation in more detail. A person is entering the FOV of C^2 from the left side. There are two persons

visible in C^1 at this instant. Both these persons are being tracked and we have a bounding box around them. By looking at the bottom part of the bounding box, we can determine quite easily that person towards the middle of the image has entered the FOV of C^2. The line that helped us determine this is L^{21}_l. The new person in C^2 is therefore assigned the same label as the one it was assigned in C^1. Note that we are considering only the left and right edges of FOV in this formulation, which is sufficient for cameras mounted at a low angle of depression. However, there is nothing in this analysis which prevents it from being extended to considering all four limits of the camera footprint, which will be necessary for images shot at a high angle of depression.

In the examples given above, it is assumed that when a person is entering the FOV of a camera, he is always visible in the FOV of another camera. This is not always the case. The person may be either entering from a door (in which case he might just "appear" in the middle of the image) or he might be entering the FOV from a point which is not visible in any other camera. If the environment is completely covered by cameras, then the latter case will never happen. However, to keep our formulation general and to not put the constraint of complete coverage of the environment, we have to consider the second case too. For example, in out-door environments, complete coverage of all parts of the site might not be possible.

Consider the scenario when a person is entering the FOV of C^i. Whether this person is visible in any other camera (C^j, $j \neq i$) or not can be determined by looking at all the FOV lines that are of the form L^{ji}_x, i.e. edge of FOV lines of other cameras as visible in this camera. These lines partition the image in C^i into (possibly overlapping) regions, marking the areas of image in C^i that correspond to FOV of other cameras. Thus we will be able to determine all the other cameras in which the current person is visible, by determining the region in which the person's feet are.

Thus, with each edge of FOV line, we also want to store an additional variable which tells us which side of this line falls in the FOV and which side falls outside it. With each line L^{ij}_x, we store a variable δ^{ij}_x. The value of this variable can be either +1 or -1, depending on which side of the line falls inside the FOV of C^i. Then, given an arbitrary point (x', y') in C^i, we can find whether it is visible in C^j or not just by determining if this point is on the correct side of *both* L^{ji}_l and L^{ji}_r. Thus, if L^{ji}_l is represented by $Ax + By + C = 0$, then let the value $L^{ji}_l(x', y')$ denote the value of the expression $Ax' + By' + C$. The point (x', y') is visible in C^j if and only if

$$\mathrm{sgn}(L^{ji}_l(x', y')) = \delta^{ji}_l \quad \text{and} \quad \mathrm{sgn}(L^{ji}_r(x', y')) = \delta^{ji}_r \qquad (1)$$

In the case when only one of the left or right lines of C^j is visible in C^i, the condition in Eq. 1 is simplified to only one of the anded terms.

When a person enters the FOV of a new camera, we first determine whether this person is visible in the FOV of some other camera or not. If the person appears in the middle of the frame, we conclude that there must be a door location at this point[1]. If the person is entering from the side of the image, then we can determine all the other cameras in which this person will also be visible (by using Eq. 1). If there is no such camera, then we assign a new label to this person. Otherwise, we need to find the previous track of the person, so that a link can be established between the two views. Say that the person entered from the left side of C^1. Then, we will search the persons visible in all cameras C^j such that $i=1, j$ satisfy Eq. 1. In all such cameras, we find which person is closest to the left edge of FOV line of C^1 in that camera. These two views will then be linked together by entering them in an equivalence table. In general, if we observe a person entering C^i from side x, then the label assigned to the new view will be:

$$label = \arg\min_k \left(D\left(P_k^j, L_x^{ij}\right) \right)$$

(2)

for all k, j such that i, j satisfy Eq.1

where k is the set of persons visible in C^j, P_k^j is the label assigned to a person in C^j and $D(P, L)$ returns the absolute distance between the center of the bottom line of the rectangular bounding box of person P and the line L. The complete algorithm for ambiguity resolution of new views is given in the inset.

3. Automatic determination of FOV lines

At the beginning when tracking is initiated, there is no information provided about the FOV lines of the cameras. The system can, however, find this information by observing motion in the environment. Suppose that there is only one person in the room. Then, when this person enters the FOV of a new camera, we find one constraint on the associated line. Two such constraints will define the line, and all constraints after that can be used in a least squares formulation. This concept is visually described in Figure 4. Once such a line is determined, it can be used to resolve ambiguous cases as described in the previous section.

This calibration is done automatically, even if the number of persons in the room in the beginning is more than one. If ambiguity exists before the lines are

---

[1] Estimates of door locations are accumulated over time so that we have a higher confidence at a location where multiple observations indicate a door

```
Repeat for each frame:

For each camera C^i
  If new person appears in the
      middle of the image, assign
      him unique label,
  Else (person appears from side x)
    Find S = set of all cameras C^j
      such that current person is
      visible in C^j (Eq 1)
    If S = φ
      Assign current person unique
      label
    Else
      For each camera C^j, j ∈ S
      For each person k in C^j
          d(j,k) = D(P_k^j, L_x^{ij})
      end
      end
    end
  end
end
Let s = row of min element of d
Let t = col of min element of d

Then P_t in C^s is the same as new person
in C^i
```

computed, then such cases are discarded. However, all correspondences that have only one possible solution contribute an additional constraint for a particular line. Thus if quick self-calibration is desired, only one person should walk around the room a few times, and this should be sufficient for determining the relationship between the cameras.

For cluttered situations where it is hard to find the correspondences to be used for calibration, we propose to use every pair of person views seen as a possible correspondence. The correct correspondences will yield a line in the single orientation, whereas the wrong correspondences will yield lines in scattered orientations. Hough transform can then be used to find the line with the maximum number of votes.

3.1 Confidence Measure:

At each time instant, the current best estimates of the FOV lines can have an associated confidence measure, based on the number of points that we have available to generate the line and the spread of these points. If two points are used to generate an edge of FOV line that are very close together in the image plane, then the line computed has a large error associated with it. Similarly the estimate for the line gets refined, as more points are available to compute the least-squares fit. We define the confidence associated with the line as a product of the

Camera 1

Camera 2

Camera 3

frame 26   frame 85   frame 385   Recovered Edge of FOV Lines

Figure 5: Determination of Edge of FOV lines using a short sequence of person walking in the room. The first 3 columns show triplets of sample images taken at same time instant. The last column shows the recovered lines

variance in spatial location of the points generating the line and the number of points contributing to it. This confidence measure increases as more points that are well spread out become available.

4. Experiments and results

To verify this formulation, we setup 3 cameras in room to cover most of the floor area. The setup is shown in Figure 4. To track persons, we used a simple background difference tracker. Each image was subtracted from a background image, and the result thresholded, to generate a binary mask of the foreground objects. We performed noise cleaning heuristically, by dilating and eroding the mask, eliminating very small components and merging components likely to belong to the same person. Occlusion is frequent in indoor environments, and to deal with occluding cases, we incorporated constant-velocity-based reasoning in our tracker. When foreground regions merge into one another, their velocities are used to predict their positions over future frames, till they separate out again. The blobs are then reassigned their labels based on minimum distance from their predicted positions. Our tracker could not deal with one case of occlusion where a person exited from the image and at the same time another person entered the image from the same location. Since the emphasis of this paper is not to develop a robust technique for tracking during person to person occlusion, but rather to demonstrate the solution to the handoff problem, we manually corrected this case of wrong tracking for the purposes of our experiments. Other than this one case, tracking was done automatically.

To determine the FOV lines initially, we had one person walk around the room briefly. All significant edge

of field of view lines were recovered from a short sequence of a single person walking in the room for only about 40 sec. Figure 5 shows some sample frames from this sequence and the edge of FOV lines recovered from this step. For the purpose of this experiment, we did not incrementally improve our estimates of lines over time. The lines found in this first step were used for the remaining experiment.

Next, two persons entered the room, walked among the cameras and exited. The tracking module tracked each view of these persons separately and assigned a unique label to each track in every camera. Overall, 10 different tracks of these persons were seen in the three cameras. Figure 6 shows all the tracks, which are 4 in C^1, 4 in C^2 and 2 in C^3. Since the actual number of persons in the world is 2, eight links need to be established between these 10 tracks. The images in Figure 7 show the 8 triplets of images that were used to establish the equivalence links between the tracks. In each of these triplets, a person is entering a new camera. The distance of all other persons from the edge of FOV of that camera is used to find the previous view of the person. The associated edge of FOV lines that are important in each case are also marked in Figure 7. The arrows in Figure 6 show the equivalence relations found out by our system. Once the arrows are marked, the complete history of tracking of the person is recovered, by linking all the tracks of the same person together. The two different colors in Figure 6 show the globally consistent labels of the two persons. It can be seen from these two figures that all handoffs were handled correctly, and the global tracking information was consistent at all times. The whole analysis part is very fast, as only the information about bounding boxes of the images and the lines is used in establishing the equivalence between tracks.

117

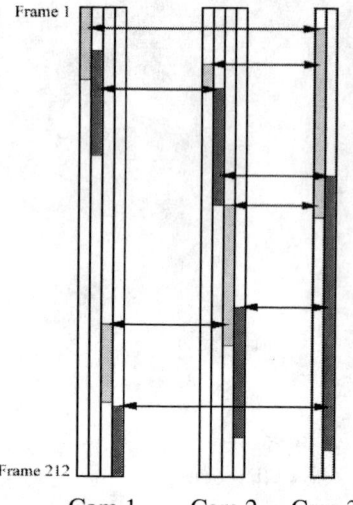

Cam 1　　Cam 2　　Cam 3

Figure 6: Tracks of two persons as seen in the three cameras. A total of 10 tracks are seen. The first two tracks in Cam 1 are new persons entering from the door. For all other tracks, an equivalence relation is established automatically, shown by the arrows. Because of the equivalence relations, globally correct labeling is achieved, shown by the different colors of the tracks.

5. Conclusion

We have described a framework to solve the handoff problem for environments that are covered by multiple cameras. We contend that the problem can be solved without going through the lengthy process of camera calibration, by finding the limits of FOV of a camera in other cameras. We outline a process to automatically find the lines representing these limits, and then use them to resolve the ambiguity between tracks. This approach does not require feature matching, which is difficult in widely separated cameras. The whole approach is simple and fast. We show results for a three-camera setup and resolve the handoff problem correctly.

6. References

[1]　P. H. Kelly, *et. al.*, "An architecture for multiple perspective interactive video", *Proc. ACM Conf. Multimedia*, pp. 201-212, 1995

[2]　Q. Cai, J. K. Aggarwal, "Tracking Human Motion in Structured Environments Using a Distributed-Camera System", *PAMI*, Vol. 2, No. 11, pp. 1241-1247, Nov 1999

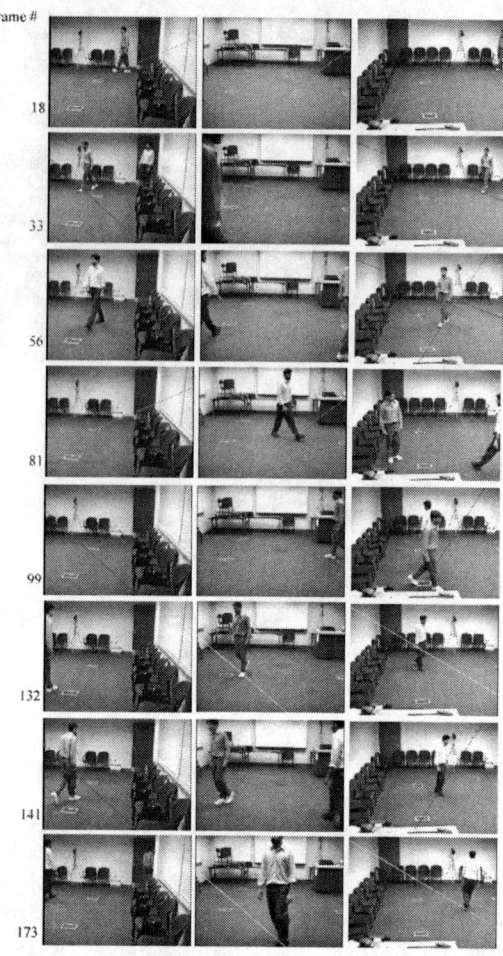

Figure 7: Handoff situations in all three cameras. Each row represents a triplet of images captured at the same time instant. One of the cameras has a person entering in it. The associated FOV lines in the other cameras help disambiguate the label of the new person. These frames result in the equivalence relationships shown in Figure 6.

[3]　Gideon P. Stein, "Tracking from Multiple View-Points: Self-calibration of Space and Time", *DARPA IUW*, Montery CA, Nov 1998

[4]　T-H. Chang, *et.al.*, "Tracking Multiple People under Occlusion using Multiple Cameras", in *BMVC*, Bristol, UK, Sept 2000

[5]　Vera Kettnaker, Ramin Zabih, "Bayesian Multi-Camera Surveillance", *CVPR*, Fort Collins, CO, June 23-25, 1999, pp. 253-259

[6]　Hanna Pasula, Stuart Russell, Michael Ostland, Ya'acov Ritov, "Tracking Many Objects with Many Sensors" In *IJCAI-99*, Stockholm 1999

Modeling/Counting

Modeling the Constraints of Human Hand Motion

John Lin, Ying Wu, Thomas S. Huang
Beckman Institute
University of Illinois at Urbana-Champaign
Urbana, IL 61801
{jy-lin, yingwu, huang}@ifp.uiuc.edu

Abstract

Hand motion capturing is one of the most important parts of gesture interfaces. Many current approaches to this task generally involve a formidable nonlinear optimization problem in a large search space. Motion capturing can be achieved more cost-efficiently when considering the motion constraints of a hand. Although some constraints can be represented as equalities or inequalities, there exist many constraints, which cannot be explicitly represented. In this paper, we propose a learning approach to model the hand configuration space directly. The redundancy of the configuration space can be eliminated by finding a lower-dimensional subspace of the original space. Finger motion is modeled in this subspace based on the linear behavior observed in the real motion data collected by a CyberGlove. Employing the constrained motion model, we are able to efficiently capture finger motion from video inputs. Several experiments show that our proposed model is helpful for capturing articulated motion.

1 Introduction

In recent years, there has been a significant effort devoted to gesture recognition and related work in body motion analysis due to interest in a more natural and immersive Human Computer Interaction (HCI). As the cost for more powerful computers decreases and PCs become more popular, a more natural interface is desired rather than the traditional input devices such as mouse and keyboard. Using gestures, as one of the most natural ways humans communicate with each other, thus becomes an apparent choice for a more natural interface [3, 8]. An effective recognition of hand gestures will provide major advantages not only in virtual environments and other HCI applications, but also in areas such as teleconferencing, surveillance, and human animation.

Recognizing hand gestures, however, involves capturing the motion of a highly articulated human hand with roughly 30 degrees of freedom (DoF). Hand motion capturing involves finding the global hand movement and local finger motion such that the hand posture can be recovered. One possible way to analyze hand motion is the appearance-based approach, which emphasizes the analysis of hand shapes in images [4, 8]. However, local hand motion is very hard to estimate by this means. Another possible way is the model-based approach [1, 2, 6, 7, 10, 13, 15]. With a single calibrated camera, local hand motion parameters can be estimated by fitting a 3D hand model to the observation images.

One method of model-based approaches is to use gradient-based constrained nonlinear programming techniques to estimate the global and local hand motion simultaneously [10]. The drawback of this approach is that the optimization is often trapped in local minima. Another idea is to model the surface of the hand and estimate hand configurations using the ''analysis-by-synthesis'' approach [6]. Candidate 3D models are projected to the image plane and the best match is found with respect to some similarity measurement. Essentially, it is a search problem in a very high dimensional space that makes this method computational intensive. A decomposition method is also proposed to analyze articulated hand motion by separating hand motion into its global motion and local finger motions [15].

Although the 3D model-based approach makes motion capturing from monocular images possible, it also faces some challenging difficulties. Many current methods for hand posture estimation basically involve the problem of searching for the optimal hand posture in a huge hand configuration space, due to the high DoF in hand geometry. Such a search process is computationally expensive and the optimization is prone to local minima. At the same time, many current approaches suffer from self-occlusion.

However, although the human hand is a highly articulated object, it is also highly constrained. There are dependencies among fingers and joints. Applying the motion constraints among fingers and finger joints can greatly reduce the size or dimensions of the search space, which in turn makes the estimation of hand postures more cost-efficient. Another major advantage of applying hand motion constraints is to be able to synthesize natural hand motion and produce realistic hand animation, which would be very useful to synthesize sign languages.

There has not been much done regarding the study of hand constraints other than the commonly used ones. Even though constraints would help reduce the size of the search space, too many or too complicated constraints would also add to computational complexity. Which constraints to adopt becomes an important issue. Some constraints have already been presented, studied, and used in many previous works [1, 2, 6, 7]. The common ones include the constraints of joints within the same finger, constraints of joints between fingers, and the maximum range of finger motions. All these are presented as either equalities or inequalities. However, due to the high flexibility in finger motion, there are yet more constraints that cannot be explicitly represented by equations.

In this paper we propose a learning approach to model the constraints directly from sampled data in the hand configuration space (C-Space). Each point in this hand configuration space corresponds to a set of joint angles of a hand state, which is commonly estimated in model-based approaches. Rather than studying the global hand motion, we will focus only on the analysis of local finger motions and constraints with the help of a CyberGlove developed by Virtual Technologies Inc. Moreover, we will study the constraints of hand motions that are natural and feasible to everyone.

2 Hand skeleton model

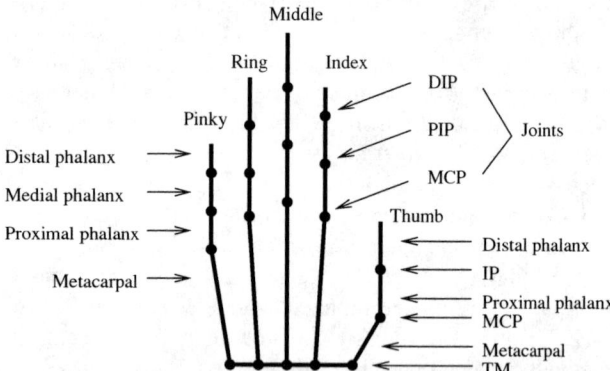

Figure 1: Kinematical structure and joint notations.

The human hand is highly articulated. To model the articulation of fingers, the kinematical structure of the hand should be modeled. In our research, the skeleton of a hand can be abstracted as a stick figure with each finger as a kinematical chain with base frame at the palm and each fingertip as the end-effector. Such a hand kinematical model is shown in Figure 1 with the names of each joint. This model has 27 Degrees of Freedom (DoF). There are 21 DoF contributed by the finger joints for the local motion and 6 DoF due to the global motion [7]. Since we

will only focus on the estimation of the local finger motions rather than the global motion, these six parameters are not considered in our current study.

Articulated local hand motion, i.e. finger motion, can be represented by a set of joint angle values. In order to capture the hand motion, glove-based devices have been developed to directly measure the joint angles and spatial positions by attaching a number of sensors to hand joints. Although the goal of vision-based hand motion analysis is to be able to recognize hand configurations without the use of attached external devices, a glove-based device will help in collecting ground truth data, which enable the modeling and learning process in visual analysis.

In our study, we employ a right-handed CyberGlove, which provides 15 sensors for measuring joint angles; therefore, we are able to characterize the local finger motion by 15 parameters. The glove can be calibrated to accurately measure the angle within 5 degrees. This is acceptable for gesture recognition; finger postures that are five degrees different would still appear to be the same posture.

3 Modeling the constraints

3.1 Constraints overview

Hand/finger motion is constrained so that the hand cannot make arbitrary gestures. There are many examples of such constraints. For instance, fingers cannot bend backward too much and the pinky finger cannot be bent without bending the ring finger. The natural movements of human hands are implicitly defined by such motion constraints.

Some motion constraints may have a closed form representation, and they are often employed in current research of animation and visual motion capturing [1, 2, 6, 7, 15]. However, many motion constraints are very difficult to express in closed forms. How to model such constraints still needs further investigation. Here we present three types of motion constraints and explain how we are able to represent hand motion with only 15 parameters instead of 21.

Hand constraints can be roughly divided into three types. Type I constraints are the limits of finger motions as a result of hand anatomy, which are usually referred to as static constraints. Type II constraints are the limits imposed on joints during motion, which are usually referred to as dynamic constraints in previous work. Type III constraints are applied in performing natural motion, and have not yet been explored. Below we will describe each type in more detail.

Type I constraints. This type of constraint refers to the limits of the range of finger motions as a result of hand anatomy. We will only consider the range of motion of each finger that can be achieved without applying external forces, such as bending fingers backward using the other hand. This type of constraint is usually represented by the following inequalities:

$$0° \le \theta_{MCP\_F} \le 90°,$$
$$0° \le \theta_{PIP\_F} \le 110°,$$
$$0° \le \theta_{DIP\_F} \le 90°, \text{ and}$$
$$-15° \le \theta_{MCP\_AA} \le 15°. \qquad (1)$$

where the subscript *F* denotes flexion and *AA* denotes abduction/adduction.

Another commonly adopted constraint states that the middle finger displays little abduction/adduction motion. The following approximation is made for the middle finger:

$$\theta_{MCP\_AA} = 0°. \qquad (2)$$

This will reduce 1 DoF from the 21 DoF model.

Similarly, the TM joint also displays limited abduction motion and will be approximated by 0 as well.

$$\theta_{TM\_AA} = 0°. \qquad (3)$$

As a result, the thumb motion will be characterized by four parameters instead of five.

Finally, the index, middle, ring, and little fingers are planar manipulators. In other words, the DIP, PIP and MCP joint of each finger move in one plane since the DIP and PIP joints only have 1 DoF for flexion.

Type II constraints. This type of constraint refers to the limits imposed on joints during finger motions. These constraints are often called dynamic constraints and can be subdivided into intrafinger and interfinger constraints. The intrafinger constraints are the constraints between joints of the same finger. A commonly used one based on hand anatomy states that for the index, middle, ring and little fingers, in order to bend the DIP joints, the corresponding PIP joints must also be bent. The relations can be approximated as follows:

$$\theta_{DIP} = \frac{2}{3}\theta_{PIP}. \qquad (4)$$

By combining Eqs. (2)-(4), we are able to reduce the model with 21 DoF to one that is approximated by 15 DoF. Experiments in previous work have shown that postures can be estimated using these constraints without severe degradation in performance.

Interfinger constraints are those imposed on joints between adjacent fingers. For instance, when an index MCP joint is bent, the middle MCP joint is forced to bend as well. Lee and Kunii [7] have performed measurements on several people and obtained a set of inequalities that approximates the limits of adjacent MCP joints. However, there are yet more constraints that cannot be explicitly represented in equations.

Type III constraints. These constraints are imposed by the naturalness of hand motions and are more subtle to detect and quantify. Almost nothing has been done to account for these constraints in simulating natural hand motion. Type III constraints differ from Type II in that

they have nothing to do with limitations imposed by hand anatomy, but rather are results of common and natural movements. For instance, the most natural way for every person to make a fist from an open hand would be to curl all the fingers at the same time instead of curling one finger at a time. Even though the naturalness of hand motions differs from person to person, it is broadly similar for everybody. This type of constraint also cannot be explicitly represented by equations.

3.2 Modeling the constraints in C-space

It is difficult to explicitly represent the constraints of natural hand motions in closed form. However, they can be learned from a large and representative set of training samples; therefore, we propose to construct the configuration space (i.e., joint angle space) and learn the constraints directly from empirical data using the approach described below. For notational convenience, let us denote the feasible C-space by $\Phi \subset \mathfrak{R}^{15}$ with each configuration denoted by ϕ.

1. *Locating base states ζ_i in* Φ. We will directly locate the base states by fixing the hand in desired configurations and measuring the 15 parameters associated with the corresponding state. Since the sensors are very sensitive to finger movements, little variations in finger postures will also be recorded and will be considered as the same state. As a result, we will use the centroid from the set of N training data $D_i = \{x_{ij}, j = 1...N\}$ as the location of the base state ζ_i. Another alternative would be to collect a huge set of training samples x_i from predefined motions and apply a clustering algorithm in order to locate the base states. This approach was taken in [11] for body posture estimation. However, since we have full control of how a hand must be configured to form the base state, we do not need to apply clustering algorithms to locate the base states in C-space.

In our model, the hand gestures are roughly classified into 32 discrete states by quantizing each finger into one of two states: fully extended or curled. The reason for choosing these two states is that the entire motion of a finger falls roughly between these two states. Therefore, the whole set of 32 states will roughly characterize the entire hand motion (Figure 2a and 2b). However, since not everyone is able to bend the pinky unless the ring finger is also bent or an external force is applied, four of the states will not be achievable by everyone without applying external forces. Therefore, these four states (Figure 2b) are not included in our set of base states in C-space modeling. Finally, the configurations that are similar are considered as the same state. For instance, the cases with five fingers spreading apart and with all fingers straightened but closed together are considered the same.

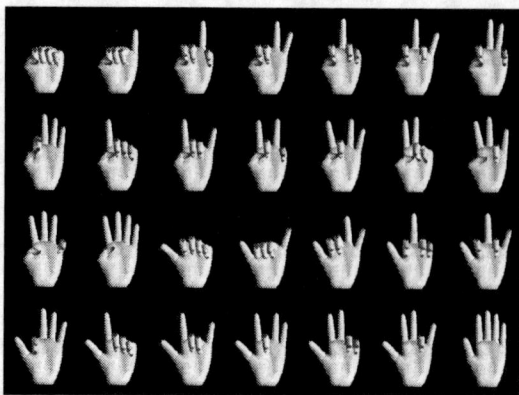

Figure 2a: Feasible base states.

Figure 2b: Infeasible base states.

2. *Motion modeling.* With the set of base states ζ_i established, we then collect motion data for state transitions in order to model the configurations during natural hand motions. A large number of sets of motion data are collected in order to observe the Type II and III constraints of natural hand motions. An example of measuring the motion of making and opening a fist is shown in Figure 3.

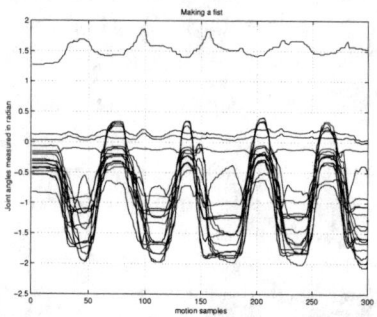

Figure 3: Joint angle measurements from the motion of making and opening a fist.

3. *Dimensionality reduction.* From Figure 3, we can clearly observe some correlations in the joint angle measurements. Therefore, together with the data collected from static states and the finger motions, we then perform principal components analysis (PCA) to reduce the dimension of the model and thus reduce the search space while preserving the components with the highest energy. We note that 95% of the energy is contained in the seven dimensions that have the largest eigenvalues. We thus perform the mapping $\Re^{15} \rightarrow \Re^7$ on Φ by projecting the original model onto a lower-dimensional subspace $\Phi^c \subset \Re^7$ with principle directions associated with these seven largest eigenvalues.

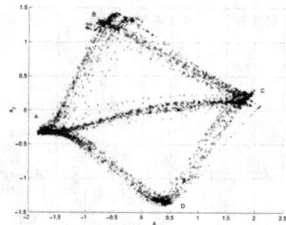

Figure 4a: Motion transitions between four states.

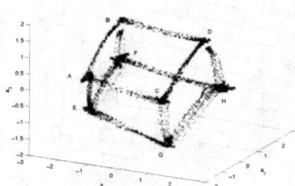

Figure 4b: Motion transitions between eight states.

4. *Interpolation in compressed C-space.* An interesting phenomenon regarding the Type III motion constraints is observed from the motion data. We observe a nearly linear transition between states in C-space. An example is shown for the case of transitioning between four states in the movements of the index and middle fingers (Figure 4a). Since only a few joints are involved in making these movements, we are able to perform PCA and project the C-space into $\Re^2$ for observation without losing much information. The four corners are the locations of the four discrete base states. A linear transition is clearly demonstrated in Figure 4a. Another example is shown in Figure 4b with three-finger motions projected onto $\Re^3$. The eight base states are roughly located at the eight corners of a cube.

Based on this observation of linear behavior, once a set of base states ζ_i has been determined, the whole feasible configuration space Φ can be approximated by these base states ζ_i and an interpolation scheme. Our approach takes a linear interpolation in the lower-dimensional configuration subspace Φ^c. For each configuration $\phi^c \subset \Phi^c$ we will represent its parameters by a polynomial interpolation, i.e.,

$$\phi^c = \sum_{i=1}^{28} \alpha_i \zeta_i^c , \qquad (5)$$

in which ζ_i^c is the projected location of base state ζ_i and

$$\sum_{i=1}^{28} \alpha_i = 1 . \qquad (6)$$

3.3 Model characteristics

Our model has three main characteristics that will help reduce the search space in gesture recognition. First, the model is compact due to the dimensionality reduction by PCA. This property also helps to compactly encode gesture representations. To obtain the data in original C-space only requires linear computations with low complexity.

Second, the motion constraints are automatically incorporated in the model. The reason for incorporating motion constraints is that we sample directly from natural hand motions. Type I constraints are represented as the boundary in the C-space, since configurations that are outside of the range permitted by hand anatomy will not be achievable in natural hand motions. Type II constraints are shown through the direction of the paths during motion. Type III constraints are observed as the straight lines from state transition paths involving multiple fingers moving together.

The third characteristic of the model is the linear behavior observed in the state transitions in the C-space. As stated before, this is the result of the Type III constraints. This observation allows us to justify the representation of all feasible configurations using linear interpolations as in Eq. (5). Furthermore, we are able to produce synthetic hand motions that replicate real hand motions with simple computations by knowing the trajectories of the state transitions. Although many current techniques exist that strive to generate lifelike hand motions [5, 9, 12, 14], many of them suffer from great computational complexity. Finally, by including time domain knowledge of the hand configuration with this linear behavior, we will be able to better predict the new configurations.

4 Posture estimations

Using the result we observe from the linear behavior, we are able to utilize the model for applications in posture estimation by taking the general approach as follows:

1. In the training stage, first associate each base state ζ_i^c with a feature vector ψ_i.

2. Extract features ψ_{input} from the input 2D image, such as edge, area, centroid, etc.

3. Compute $\alpha_i = h(\psi_i, \psi_{input})$, where $h(\psi_i, \psi_{input})$ measures the closeness of ψ_{input} to ψ_i, and α_i are normalized as in Eq. (6).

4. Based on the observation made from Type III motion constraints, linearly interpolate the estimated configuration in the compressed space Φ^c:

$$\phi_{estimate}^c = \sum_{i=1}^{28} \alpha_i \zeta_i^c \qquad (7)$$

5. Reconstruct the estimated configuration state $\phi_{estimate} \subset \Re^{15}$ from $\phi_{estimate}^c$.

5 Experiments

In order to evaluate the validity of this model, we perform some experiments in synthesizing realistic finger motions and estimating the postures constituted by a subset of the 28 base states. The input images are assumed to be segmented. In our current experiment, we manually identify the 2D locations of the fingertips relative to the center of the base of the palm as the features ψ_{input} for each input image.

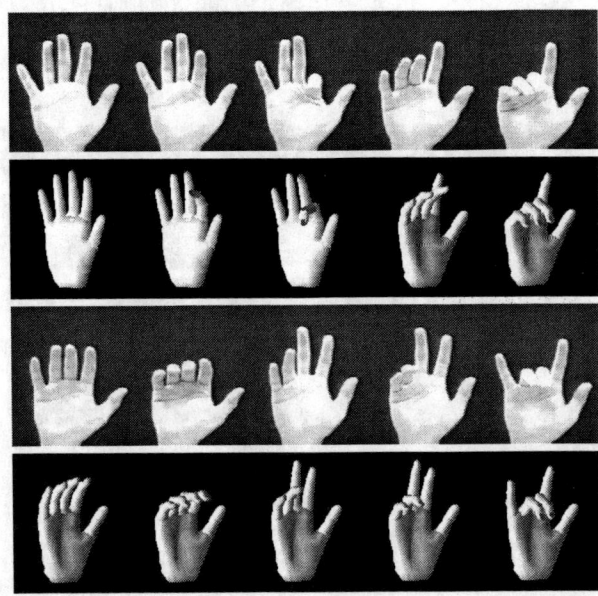

Figure 5: Configuration estimations.

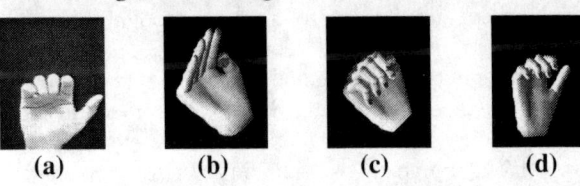

(a)   (b)   (c)   (d)

Figure 6: Comparison of different techniques. (a) original image. (b) estimation with Type I constraints only. (c) estimation with Type I & II constraints only. (d) estimation with Type I, II & III constraints.

The results of the experiments are shown in Figure 5. The first and third rows are the input images and the second and fourth rows are the corresponding reconstructed 3D hand models based on the estimation by our approach. The results are visually agreeable. Such preliminary experiments show that the motion constraints play an important role in hand posture estimation. More accurate and cost-efficient estimation can be obtained when a better motion constraint model is applied. Moreover, better results can be obtained with better feature extraction methods, which will be implemented in the future research.

A comparison of estimations using different types of constraints is also shown in Figure 6(a)-(d). In Figure 6(b), estimation with only Type I constraints results in a feasible, yet unnatural configuration. In Figure 6(c), a closer approximation is obtained by applying Type I & II constraints. Some additional adjustments are required in order to approximate the configurations correctly. Finally, applying all three types of constraints together produces the better result with a more natural approximation in Figure 6(d).

Another application is hand motion synthesis by reconstructing the sequences of configurations along the lines that approximate the state transitions (Figure 7). Since the lines are the approximations of the original real motion data, the reconstructed sequences also incorporate the constraints, which make the motion realistic.

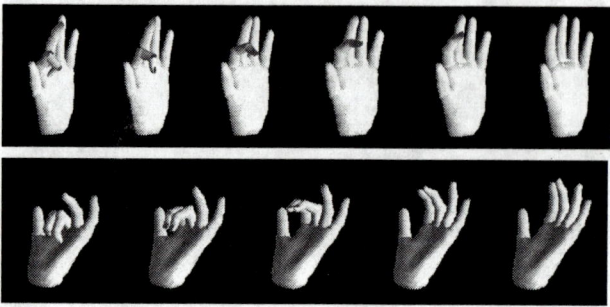

Figure 7: Sequences of synthesized finger motions.

6   Conclusion

A posture estimation problem generally involves a search in a high dimensional C-space. Useful hand constraints have been demonstrated to greatly reduce the search space, and thus improve gesture recognition results. Many constraints can be represented in simple closed forms while many more cannot and have not been found.

In this paper, we presented a novel approach to model the hand constraints. Our model has three characteristics. First, it is compact by utilizing PCA. Second, it incorporates constraints that can and cannot be represented by equations. Third, it displays a linear behavior in state transitioning as a result of natural motion. These properties together simplify configuration estimation in the C-space as shown in Eq. (5) by a simple interpolation with linear polynomials. Some preliminary gesture estimation experiments are shown, taking advantage of this model.

However, there is still much to be done to improve this model. For instance, more states can be included to further refine the model. Deciding which states to choose will require more analysis of the C-space. Furthermore, other constraints might exist in the C-space that have not yet been observed. Finally, even though a nearly linear behavior is observed in state transition, it is not exactly linear. A more detailed study can better approximate the trajectories, which in turn would help improve the configuration estimation. Nevertheless, our constraints modeling provides a different interpretation of hand motions and the current results look promising.

Acknowledgement

This work was supported in part by NSF CDA-96-24396, NSF IRI-9634618, and ARL Cooperative Agreement DAAL01-96-2-0003.

References

[1] C. Chang, W. Tsai, "Model-Based Analysis of Hand Gestures From Single Images Without Using Marked Gloves Or Attaching Marks on Hands", *ACCV2000*, 2000, pp. 923-930.

[2] C.S. Chua, H. Y. Guan and Y. K. Ho, "Model-based Finger Posture Estimation", *ACCV2000*, 2000, pp. 43-48.

[3] R. Cipolla and A. Pentland (Editors), *Computer Vision for Human-Machine Interaction*. Cambridge: Cambridge University Press, 1998.

[4] T. Heap and D. Hogg, "Wormholes in shape space: tracking through discontinuous changes in shape," in *Proc. of IEEE ICCV98*, 1998, pp. 344-349.

[5] P. Kalra, N. Magnenat-Thalmann, L. Moccozet, G. Sannier, A. Aubel, and D. Thalmann, "Real-time animation of realistic virtual humans," *IEEE Computer Graphics and Applications,* vol. 18, issue:5, pp. 42-56, Sept-Oct, 1998.

[6] J. Kuch and T. S. Huang, "Vision-Based Hand Modeling and Tracking for Virtual Teleconferencing and Telecollaboration", *ICCV95*, 1995, pp.666-671.

[7] J. Lee, T. Kunii, "Model-based Analysis of Hand Posture", *IEEE Computer Graphics and Applications*, Sept., pp. 77-86, 1995.

[8] V. Pavlovic, R. Sharma, T. S. Huang, "Visual Interpretation of Hand Gestures for Human-Computer Interaction: A Review," *IEEE PAMI*, vol. 19, No. 7, pp. 677-695, July, 1997.

[9] Z. Popović and A. Witkin, "Physically based motion transformation," *SIGGRAPH99*, 1999, pp. 11-20.

[10] J. Rheg, T. Kanade, "Model-Based Tracking of Self-Occluding Articulated Objects", *Proc. of IEEE ICCV95*, 1995, pp. 612-617.

[11] R. Rosales and S. Sclaroff, "Inferring body pose without tracking body parts," in *Proc. of IEEE CVPR*, 2000, vol. 2, pp. 721-727.

[12] C. Rose, B. Guenter, B. Bodenheimer, and M. F. Cohen, "Efficient generation of motion transitions using spacetime constraints," in *Proc. of SIGGRAPH 96*, 1996, pp. 147-154.

[13] N. Shimada, et al., "Hand Gesture Estimation and Model Refinement Using Monocular Camera-Ambiguity Limitatio by Inequalty Constraints," *Proc. of the 3rd Conf. On Face and Gesture Recognition*, 1998, pp. 268-273.

[14] A. Witkin and M. Kass, "Spacetime constraints," in *Proc. of SIGGRAPH 98*, 1998, pp. 159-168.

[15] Y. Wu, T. S. Huang, "Capturing Human Hand Motion: A Divide-and-Conquer Approach", *Proc. of IEEE ICCV99*, 1999, vol. 1, pp. 606-611.

Person Counting Using Stereo

David Beymer[*]
SRI International
333 Ravenswood Ave
Menlo Park, CA 94025
beymer@ai.sri.com

Abstract

Stores and shopping malls would like to keep track of shopper volume by employing automatic techniques for counting shoppers. Existing approaches instrument doors with infrared beams and count beam interruptions, but this approach cannot resolve groups of people well. We are applying a vision-based approach that detects and tracks people from a stereo camera mounted above a door and pointing down. After applying real-time stereo and 3D reconstruction, the system segments the scene by selecting stereo pixels falling inside a 3D volume of interest, which is placed to capture the heads and torsos of adult shoppers. The main novelties of our approach include (1) remapping the stereo disparities to an orthographic "occupancy map", which simplifies person modeling, and (2) tracking people using a Gaussian mixture model. On a test set of 900 enter/exit events in 4 hours of video, our system has achieved a net counting error rate of just 1.4%.

1 Introduction

Driven by the need for marketing information about shoppers, shopping malls and department stores are deploying more and more sensors to keep track of shoppers. Given a system that can count shoppers or track movement within a store, malls can set lease rates based on foot traffic, stores can monitor shopper volume, and the effectiveness of store displays can be evaluated.

In this paper, we focus on the problem of counting shoppers as they enter or exit a door. Allowing the doors to be double doors makes the problem difficult, as multiple people can be entering side by side, or one person can be stopped in the doorway while others pass. In addition, we want to exclude counting objects such as shopping carts or small children.

There are already a number of different types of sensing systems in place for counting shoppers. Infrared beam counters are placed so that a beam is interrupted when shoppers enter or exit. While two parallel beams can be used to give the direction of the shopper, the system will undercount two people who are side by side. Infrared motion detectors are also in use, but they suffer from similar problems with resolving groups of people.

Person tracking systems using a camera mounted overhead and pointing straight down are becoming commercially available. Orienting a camera this way alleviates the partial occlusion problem associated with more oblique mounting angles. A monocular system from ShopperTrak is being deployed by The Gap and Banana Republic. A stereo-based person tracking system from Point Grey Research, CenSys3D, was recently demonstrated at CVPR 2000.

Our approach to person counting is to track stereo blobs within a 3D volume of interest. A stereo camera is mounted in the ceiling near the door, pointing straight down. An operator first specifies a 3D volume of interest in a normally empty region by the door, but outside the door swing region. The person counter then detects and tracks the heads and upper torsos of people as they pass through the volume of interest. Because the ceilings of a typical store installation are only 8-9 feet, wide field of view lenses will be required to capture a region wide enough to monitor a double door. Since this introduces severe radial lens distortion and accentuates perspective distortion, our system remaps the stereo disparities into an orthographic, top-down view.

The main components of our system are real-time stereo, reprojection to an orthographic view, and multiple person detection and tracking. The stereo system performs real-time 3D reconstruction by employing table lookup to map from image coordinates and disparities to 3D. Based on the 3D coordinates, the scene is (1) segmented by filtering out pixels outside the volume of interest, and (2) reprojected to a top-down, orthographic view. Finally, people are detected and tracked in the orthographic reprojection using a Gaussian mixture model and Kalman filtering.

What are the advantages of a stereo-based system over non-video or monocular approaches to person counting? Compared to infrared beam counters, our system can handle multiple people moving side by side, avoiding the undercounting problem. Compared to monocular approaches, stereo provides a robust method for estimating object height, which is useful for distinguishing adult shoppers from children or shopping carts. Also, the coupling of 3D reconstruction and the volume of interest provides for an extremely robust

---

[*] Author's current affiliation: IBM Almaden Research Center, 650 Harry Rd, San Jose, CA 95120

127

segmentation. A popular alternative segmentation approach is background differencing, which introduces the difficult problem of maintaining a background image over time.

In this paper, we first discuss related work, and then we describe camera setup, calibration, and specification of the volume of interest. The next section describes the main tracking algorithm, including real-time stereo and reprojection. Next, we present experimental results of the system on roughly four hours of video, and we close with a discussion of future work.

2 Related work

A large number of person tracking systems have been developed over the last several years, and the majority of them use color[1-5], background subtraction[3, 6-9], or contour modeling[10, 11]. In addition, a number of systems can track multiple people[3, 4, 6, 7, 9, 12], which adds the burden of maintaining person ID and handling occlusion events. In this section on related work, we concentrate on systems that either use stereo or Gaussian modeling.

The main advantages of adding stereo to a person tracking system include (1) segmentation, (2) locating objects in 3D, and (3) handling occlusion events. In [3] and [13], large baseline stereo is used to triangulate the location of person blobs in 3D. In [4], dense stereo is computed in real-time and used to segment the scene into patches of slowly varying disparity. Background subtraction on dense stereo disparities is performed in [5, 6, 12], where [6] includes image intensities and [5] includes color. [6] also uses stereo to help disambiguate occlusion events. Finally, [14] and [15] have started fitting parametric human models to dense stereo maps, with the hope that stereo can help with self-occlusion of body parts. In our use of stereo for segmentation, we reconstruct the scene in 3D and select disparity pixels in a volume of interest.

Gaussian functions have been used in person tracking systems for modeling multimodal distributions and articulated human models. [1] and [14] model human bodies as connected 3D Gaussian blobs. Gaussian mixture models have been used to model the multimodality of background images[16, 17], making background subtraction more robust to shadows or blowing leaves. In the area of multi-hypothesis tracking, [18] has used a 'piecewise' Gaussian model to represent multimodality in tracking state space, allowing the system to maintain a number of competing tracking states. This is related to Condensation tracking[11], which represents multimodality non-parametrically with a set of samples in state space. In our system, we are using Gaussian mixtures to model multiple people, with one Gaussian per person.

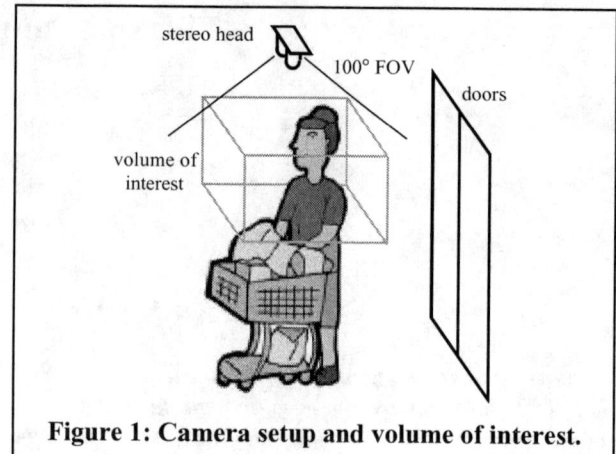

Figure 1: Camera setup and volume of interest.

3 Hardware setup

In this section, we describe the hardware setup for our person counter, including the camera setup, calibration, and specifying the volume of interest.

The stereo head we use is Videre Design's STH-V1, an inexpensive, low power stereo head with analog CMOS cameras[19]. As shown in Figure 1, the stereo head is ceiling mounted a few feet from the doors in order to minimize door visibility. Since the ceilings near the doors are typically only 8-9 feet, a wide field of view (FOV) is required to cover a set of double doors; our 2.1mm lenses give the system a horizontal FOV of approximately 100°. The choice of high FOV lenses

raw left image   undistorted, rectified, with VOI

Figure 2: Example image from left camera: (left) raw image, (right) undistorted, rectified, with VOI.

increases the effect of radial lens distortion, as shown in the example left image in Figure 2.

During stereo head calibration, we estimate the intrinsic parameters of the left and right cameras and the rigid transform between the two cameras. During the procedure, five views of a planar calibration pattern are presented to the stereo head, where the pattern is the 4x4 grid of squares used by Tsai[20]. Camera intrinsics are modeled after Tsai[20], using one radial distortion parameter κ_1, and all the parameters are estimated using a bundle adjustment procedure applied to the five stereo pairs. From the calibration, the system computes an image warp for (1) removing radial distortion, and (2) rectifying the left and right images to align epipolar lines for stereo matching (see Figure 2).

The power of a calibrated stereo head is the ability to perform 3D reconstruction. In our person counting setup, this allows us to segment the scene by focusing on objects falling inside the 3D volume of interest (VOI). Proper placement of the VOI by a system operator allows the system to focus on the head and upper torso of adult shoppers, ignoring shopping carts and small children. The VOI is currently a 3D box (see Figures 1 and 2), although the shape can be easily customized, if necessary. Reprojection of stereo disparities into an orthographic view will be performed inside the VOI.

4 Tracking in occupancy maps

As mentioned in the introduction, tracking is done using stereo primarily because the segmentation problem is simplified. Beyond using stereo, though, we remap disparities to what we call an *occupancy map*, an orthographic reprojection of the scene disparities inside the VOI. In this section, we discuss the creation of occupancy maps and our use of Gaussian mixture models to track people in them.

4.1 Occupancy maps

Perspective distortion tends to elongate stereo blobs near the boundaries of the image, as people present more of a side view to the camera. As shown in Figure 3a, the elongation is towards the optical center of the camera. If we can resample the image from a vertical projection, then the stereo blobs should be more compact and simpler to model mathematically. 3D reconstruction from stereo enables our system to perform the orthographic resampling. Furthermore, the resampled image can be computed in real-time through the use of lookup tables.

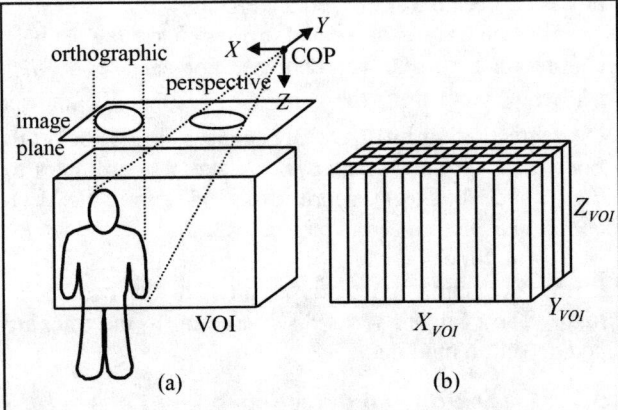

Figure 3: (a) A person's projection is more compact under orthographic projection than perspective, and (b) the division of the VOI into buckets.

We call the resampled image an *occupancy map*. Occupancy maps divide the *X-Y* plane of the VOI into a set of discrete vertical buckets, as shown in Figure 3b. At location $occ(X_{VOI}, Y_{VOI})$, we accumulate all disparity pixels that land in the corresponding bucket. We can also work forward from the disparity image to see how pixels map into the occupancy map. Given an image disparity at $(x, y, disp)$, we first reconstruct the point in 3D (X, Y, Z) and then we map to (X_{VOI}, Y_{VOI}) using a simple windowing transform on (X, Y). Finally, we increment the value of $occ(X_{VOI}, Y_{VOI})$.

Since closer objects appear larger in the image, having each disparity pixel $(x, y, disp)$ contribute a fixed amount to $occ(X_{VOI}, Y_{VOI})$ favors closer objects. Thus, a technique is needed to compensate for the depth of a pixel. Using the stereo equation $Z = bf/disp$, where b and f are the baseline and focal length, one can easily show that incrementing $occ(X_{VOI}, Y_{VOI})$ by $disp_{nom}/disp$ compensates for range, where the accumulation is 1 for pixels at disparity $disp_{nom}$. An example of an occupancy map is shown in Figure 4. We always smooth the occupancy map to compensate for image noise and stereo matching errors.

4.2 Tracking using Gaussian mixture models

Since the head and torso pixels map to Gaussian blobs in the occupancy map, we have explored using a Gaussian mixture model to approximate the occupancy map $occ(x, y)$. (In this section, we use lower case x and

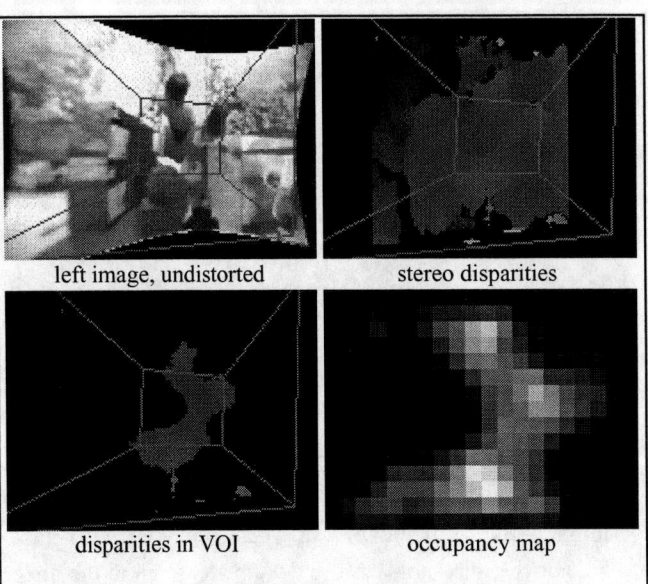

left image, undistorted   stereo disparities

disparities in VOI   occupancy map

Figure 4: Stereo disparities are first filtered using the VOI and then orthographically reprojected to produce the occupancy map.

y to index the occupancy map). Each individual being tracked will be represented by a single Gaussian in the mixture. The tracking procedure is hence driven by the function approximation of $occ(x,y)$.

Two mixture models have been explored for representing occupancy. Let the mixture model G be composed of n Gaussians G_i, where

$$G_i(x,y;x_i,y_i,s_i,\sigma_i) = s_i\, e^{-\frac{[(x-x_i)^2+(y-y_i)^2]}{\sigma_i^2}}.$$

For the sake of brevity we will often omit the parameters $x_i, y_i, s_i,$ and σ_i and write G_i as $G_i(x,y)$. We have explored:

1. *Additive mixture model*: $G(x,y) = \sum_{i=1}^{n} G_i(x,y)$

2. *Competitive mixture model*: $G(x,y) = \underset{i=1}{\overset{n}{\text{Max}}}\, G_i(x,y)$

The main difference between the models is that the competitive model explicitly incorporates segmentation of the occupancy map. That is, a Gaussian G_i (i.e. person i being tracked) claims the pixel set

$$\{(x,y)\,|\,G_i(x,y) > G_j(x,y), j \neq i\}.$$

The additive model, on the other hand, is more liberal and allows multiple Gaussians to explain occupancy at a given (x,y). We found that when two people are close to one another, this additive mixing reduced the Gaussians' ability to lock onto individual modes of $occ(x,y)$, although the overall approximation was good. The competitive model was better able to track the modes of $occ(x,y)$ for groups of people.

The "measurement" step of the tracker fits the competitive mixture model to the occupancy map by iteratively performing segmentation and parameter update steps. For n people, the measurement process is:

Iterate until Gaussian parameters settle
  For each person i:
    segment:
    $mask_i(x,y) = \{(x,y)\,|\,G_i(x,y) > G_j(x,y), j \neq i\}$
    update G_i: minimize
    $\sum_{mask_i}(occ(x,y) - G_i(x,y))^2$ using one step of
    Newton-Raphson iteration.

An example of this fitting process is shown in Figure 5.

In this scheme, the system detects new people by looking for local maxima in $(occ(x,y) - G(x,y))$ that are above a threshold. More details are given in the next section, which provides a complete description of the system.

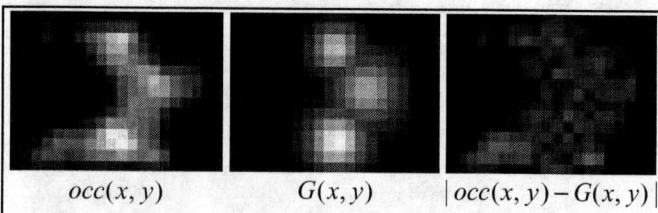

| $occ(x,y)$ | $G(x,y)$ | $|occ(x,y)-G(x,y)|$ |

Figure 5: The measurement process in the tracker approximates *occ(x,y)* with a mixture of Gaussians *G(x,y)*.

5 Person counting algorithm

The goal of the person counting system is to count the number of people entering and exiting a doorway. Figure 6 shows a flowchart of the entire system, beginning with stereo and ending with person track classification. In this section, we give the details of the overall counting system.

5.1 Stereo and occupancy map

For stereo, we use SRI's Small Vision System, a real-time, area correlation-based stereo implementation that runs on standard PC hardware [19]. See Figure 4 for an example disparity map, where lighter grey levels correspond to larger disparities and closer objects. To assure stereo operation at frame rate with extra cycles to spare for the tracker, we capture images at 160x120 and compute stereo over 16 disparities.

Segmentation of pixels inside the VOI and reprojection to form the occupancy map is performed using table lookup. For each possible x, y, and disparity $disp$, we need to store (1) an "inside VOI" segmentation bit, and (2) the coordinates (X_{VOI}, Y_{VOI}) if the pixel is indeed inside the VOI. Thus, the tables for segmentation and VOI coordinates are 3D, indexed by x, y, and $disp$. Our tables are reasonable in size (2.6 MB total), however, because of our reduced image size.

The tables are generated offline once the camera calibration and VOI are known. For each x, y, $disp$ triplet, we reconstruct the point (X,Y,Z) in 3D, and set the segmentation bit if the 3D point is inside the VOI. For interior points, we map the X and Y coordinates to the proper bucket (Figure 3b) and store the VOI coordinates in the second table.

At run time, we can easily test to see if a disparity is in the VOI and immediately map it to the occupancy map. The occupancy map is passed on to the tracking and detection modules.

5.2 Tracking and detection

As mentioned in the previous section, each person is represented by a Gaussian that tracks their movement in the occupancy map. Thus, the main part of a person's state vector is the vector $(x_i, y_i, s_i, \sigma_i)$ parameterizing

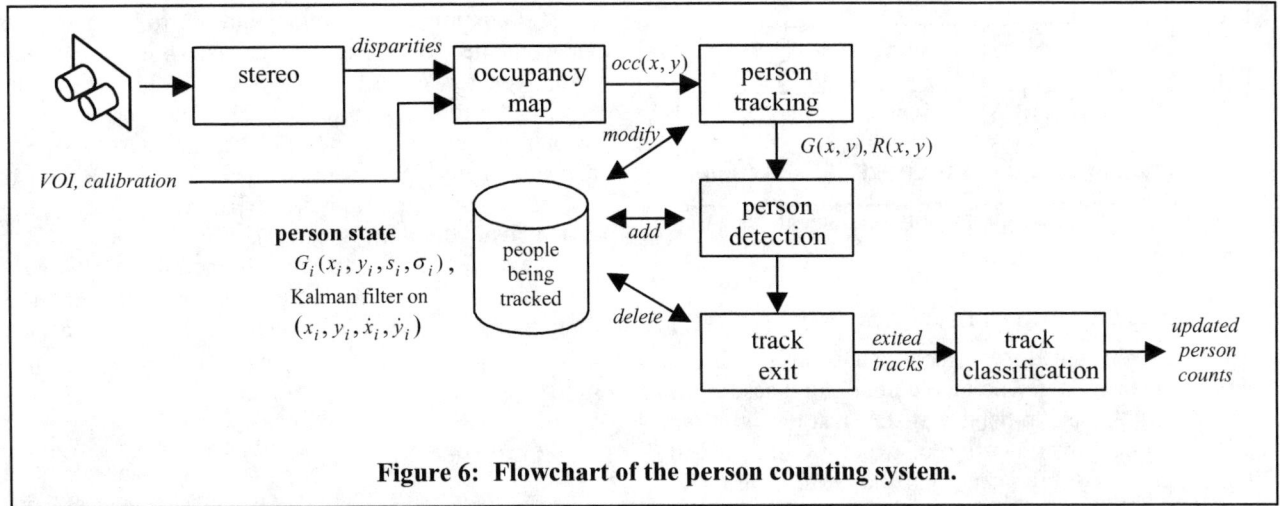

Figure 6: Flowchart of the person counting system.

the Gaussian. In addition, the person's position (x_i, y_i) is tracked with a Kalman filter; the Kalman filter maintains position and velocity $(x_i, y_i, \dot{x}_i, \dot{y}_i)$ and assumes a simple constant velocity motion model.

The tracker follows a simple Kalman filter framework with the Gaussian mixture modeling as the "measurement" process. For each person we predict the expected location (x_i, y_i) in the occupancy map; this location is used as initial conditions for the Gaussian fitting process. Next, we run the measurement process described in section 4.2 to update the Gaussian locations.

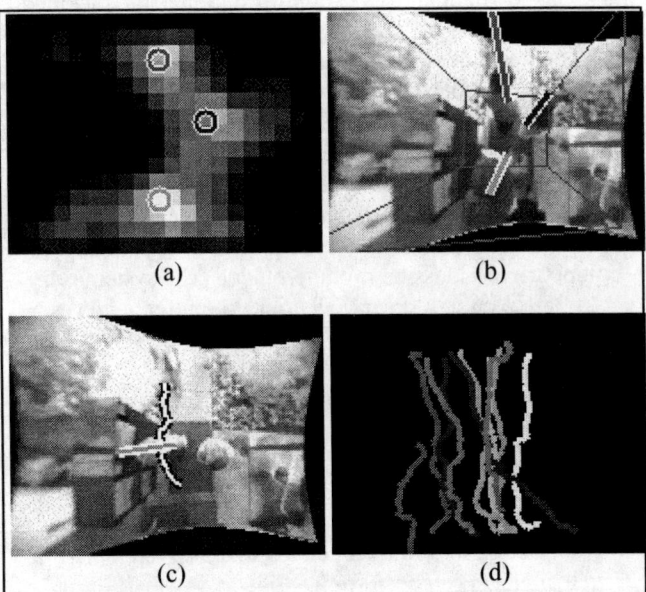

Figure 7: Tracker results: (a) the occupancy map computed from (b) with Gaussian centroids plotted as circles. In (b), person locations are plotted by projecting a 3D vertical bar. In (c), a complete track over time is shown, with a vertical bar showing the present person location. (d) shows several tracks over time in the occupancy map.

Finally, we perform a Kalman filter update on $(x_i, y_i, \dot{x}_i, \dot{y}_i)$. Figure 7 shows a tracking example, where (a) shows person locations plotted as circles in the occupancy map computed from (b). In Figure 7(b), person locations are plotted by projecting a 3D vertical bar through the 3D Gaussian centroid. In Figure 7(c), we show a track over time, where a vertical bar shows the present person location. Finally, in (d), a number of tracks are shown in the occupancy map.

The detection of new people entering the VOI uses the residual image $R(x, y) = occ(x, y) - G(x, y)$ from the tracker. $R(x, y)$ contains stereo data that is not explained by current tracks. The detection algorithm first smoothes $R(x, y)$ for noise immunity and then looks for local maxima above a detection threshold. If a local maximum is found, we fit a Gaussian to $R(x, y)$ and begin a new track. To avoid a problem with multiple detections when a person is slightly multimodal in $occ(x, y)$, we have an additional test that tries to link the new detection to an existing track. Given the new potential track and an existing track, this linkage test is based on (1) their proximity, and (2) the drop-off in occupancy between the two Gaussian modes.

5.3 Track exit and classification

When a person exits the VOI, their track is deleted by noticing either:

(1) (x_i, y_i) is out of bounds,

(2) $\sigma_i <$ threshold (Gaussian width is too narrow),

(3) $s_i <$ threshold (occupancy is too weak).

Upon track exit, the system classifies the track in order to one of four categories, as shown in Figure 8, based on a simple geometric analysis of the track path. Categories "reenter" and "passby" do not affect the person counts, while an "enter" or "exit" event will increment the corresponding count. Tracks will be rejected if they are

131

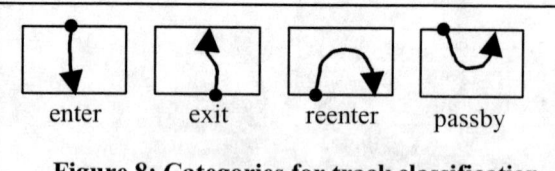

| enter | exit | reenter | passby |

Figure 8: Categories for track classification.

too short or if they do not fit in one of the four categories.

6 Experimental results

We have evaluated the person counter on roughly 4 hours of video collected at double door entrances on the SRI campus. About 3.5 hours were taken at lunchtime at the SRI cafeteria over a period of two days, and around 30 minutes was taken at a SRI auditorium during a meeting. The data was recorded to VCR (our stereo head interlaces the left and right views into a single video feed) and digitized into clips lasting from a few seconds to 20 seconds. Clips are chosen so that the track classifications are unambiguous. Finally, for each clip we manually determine ground truth counts; for the ith clip let the count for each event type be $enter_i^{gt}$, $exit_i^{gt}$, $reenter_i^{gt}$, and $passby_i^{gt}$. The net count for a clip i is $net_i^{gt} = enter_i^{gt} - exit_i^{gt}$.

Next, the person counter is run on all the clips, recording the tracker counts similarly as $enter_i^{tr}$, $exit_i^{tr}$, $reenter_i^{tr}$, and $passby_i^{tr}$. Three error rates are defined:

(1) Net count error: $error_{NC} = \dfrac{\sum_i |net_i^{gt} - net_i^{tr}|}{\sum_i (enter_i^{gt} + exit_i^{gt})}$

(2) Enter error: $error_{enter} = \dfrac{\sum_i |enter_i^{gt} - enter_i^{tr}|}{\sum_i enter_i^{gt}}$

(3) Exit error: $error_{exit} = \dfrac{\sum_i |exit_i^{gt} - exit_i^{tr}|}{\sum_i exit_i^{gt}}$

Table 1 shows our error rates for three groups of clips, as well as the total error rate. The most important number is in the lower left: 13 net count errors for 905 enter/exit events. The two dominant sources of error are (1) tall people being "multimodal" in the occupancy map and hence overcounted, and (2) the track classifier rejecting a track because of a wild motion at track exit.

	Error rates		
	Net count	Enter	Exit
Cafe1	2.1% (10/471)	2.1% (5/241)	1.8% (4/218)
Cafe2	0.9% (3/305)	2.1% (4/193)	1.9% (3/159)
All hands	0.0% (0/79)	0.0% (0/69)	0.0% (0/4)
Total	**1.4% (13/905)**	**1.8% (9/503)**	**1.8% (7/381)**

Table 1: In our experimental results, the system achieved a net count error of 1.4% on 905 enter/exit events.

It is possible for the net count error to be lower than the combined enter and exit errors, as enter and exit errors can cancel each other within a clip. Also, the enter/exit error rates do not include clips with passby or reenter events since how the tracker classifies them depends on the placement of the VOI (i.e. the path of such an event can avoid the VOI; enter or exit events cannot avoid the VOI, however).

The system has achieved a frame rate of 20-25 Hz, depending on the number of people being tracked. This timing was done on a dual Pentium II running at 400 MHz (but the system is single threaded and only uses one processor).

7 Conclusion

The main contributions of this paper are the introduction of the occupancy map and the use of Gaussian mixture models for tracking multiple people. The occupancy map remaps stereo disparities in a way that simplifies camera viewpoint and subsequent modeling. In our person counting problem, it turned the problem of tracking irregularly shaped blobs with varying depth into one of tracking Gaussian shaped blobs. Second, mixture models have previously been applied for modeling multimodal distributions (e.g. background modeling) and handling ambiguous state spaces (e.g. multi-hypothesis tracking). We have applied it to the occupancy map, where the modes correspond to people, showing that it is useful for multiperson tracking.

With regards to the person counting problem, we have demonstrated a real-time system that can resolve groups of people, unlike the popular infrared beam systems commonly in use today. Our stereo-based system segments the scene by focusing on pixels in a 3D volume of interest and remaps those pixels to an orthographic view called the occupancy map. Persons are tracked in the occupancy map by using a "competitive" Gaussian mixture model. The system runs at 20-25 Hz on a 400 MHz Pentium II, and it achieves a net counting error of only 1.4% on a test set of 900 enter/exit events.

For future work, we plan to perform more field testing, especially at department stores or shopping malls. The error rate could be reduced by investigating the problem of tall people appearing to be multimodal in the occupancy map. Finally, we would like to simplify the camera calibration procedure to the point where a field technician can perform it.

8 References

[1] Wren, C., et al., *Pfinder: Real-Time Tracking of the Human Body.* IEEE Transactions of Pattern Analysis and Machine Intelligence, 1997. 19(7): p. 780-785.

[2] Fieguth, P. and D. Terzopoulos. *Color-Based Tracking of Heads and Other Mobile Objects at Video Frame Rates.* in *IEEE Conference on Computer Vision and Pattern Recognition.* 1997. San Juan, Puerto Rico, pp. 21-27.

[3] Rehg, J.M., M. Loughlin, and K. Waters. *Vision for a Smart Kiosk.* in *IEEE Conference on Computer Vision and Pattern Recognition.* 1997. San Juan, Puerto Rico, pp. 690-696.

[4] Darrell, T., G. Gordon, and M. Harville. *Integrated person tracking using stereo, color, and pattern detection.* in *IEEE Conference on Computer Vision and Pattern Recognition.* 1998. Santa Barbara, California, pp. 601-608.

[5] Krumm, J., et al. *Multi-Camera Multi-Person Tracking for EasyLiving.* in *IEEE International Workshop on Visual Surveillance.* 2000. Dublin, Ireland.

[6] Haritaoglu, I., D. Harwood, and L.S. Davis. W^4S: *A Real-Time System for Detecting and Tracking People in 2 1/2D.* in *European Conference on Computer Vision.* 1998. Freiburg, Germany, pp. 877-892.

[7] Kanade, T., et al. *Advances in Cooperative Multi-Sensor Video Surveillance.* in *Proceedings of the Image Understanding Workshop.* 1998. Monterey, California, pp. 3-24.

[8] Boult, T.E., et al. *Frame-Rate Omnidirectional Surveillance and Tracking of Camoflaged and Occluded Targets.* in *IEEE International Workshop on Visual Surveillance.* 1999. Fort Collins, CO, pp. 48-55.

[9] Rosales, R. and S. Sclaroff. *3D Trajectory Recovery for Tracking Multiple Objects and Trajectory Guided Recognition of Actions.* in *IEEE Conference on Computer Vision and Pattern Recognition.* 1999. Fort Collins, CO, pp. 117-123.

[10] Baumberg, A. and D. Hogg. *Learning Flexible Models from Image Sequences.* in *European Conference on Computer Vision.* 1994. Stockholm, Sweden, pp. 299-308.

[11] Isard, M. and A. Blake. *Contour Tracking by Stochastic Propagation of Conditional Density.* in *European Conference on Computer Vision.* 1996, pp. 343-356.

[12] Beymer, D. and K. Konolige. *Real-Time Tracking of Multiple People Using Stereo.* in *IEEE Frame Rate Workshop.* 1999. Corfu, Greece.

[13] Azarbayejani, A., C. Wren, and A. Pentland, *Real-Time 3-D Tracking of the Human Body,* MIT Media Lab, Technical Report No. 374, 1996

[14] Jojic, N., M. Turk, and T.S. Huang. *Tracking Self-Occluding Articulated Objects in Dense Disparity Maps.* in *IEEE International Conference on Computer Vision.* 1999. Kerkyra, Greece, pp. 123-130.

[15] Lin, M.H. *Tracking Articulated Objects in Real-Time Range Image Sequences.* in *IEEE International Conference on Computer Vision.* 1999. Kerkyra, Greece, pp. 648-653.

[16] Grimson, W.E.L., et al. *Using Adaptive Tracking to Classify and Monitor Activities in a Site.* in *IEEE Conference on Computer Vision and Pattern Recognition.* 1998. Santa Barbara, California, pp. 22-29.

[17] Friedman, N. and S. Russell. *Image Segmentation in Video Sequences: A Probabilistic Approach.* in *Proceedings of the Thirteenth Conference on Uncertainty in Artificial Intelligence.* 1997. Providence, RI: Morgan Kaufmann.

[18] Cham, T.-J. and J.M. Rehg. *A Multiple Hypothesis Approach to Figure Tracking.* in *IEEE Conference on Computer Vision and Pattern Recognition.* 1999. Fort Collins, CO, pp. 239-245.

[19] Konolige, K. *Small Vision Systems: Hardware and Implementation.* in *Eighth International Symposium on Robotics Research.* 1997. Hayama, Japan, pp. 111-116.

[20] Tsai, R.Y., *A Versatile Camera Calibration Technique for High-Accuracy 3D Machine Vision Metrology Using Off-the-Shelf TV Cameras and Lenses.* IEEE Journal of Robotics and Automation, 1987. **RA-3**(4): p. 323-344.

Interaction and Shape Estimation

Realistic Synthesis of Novel Human Movements from a Database of Motion Capture Examples

Luis Molina Tanco   Adrian Hilton
Centre for Vision, Speech and Signal Processing
School of Electronical Engineering, Information Technology and Maths
University of Surrey, Guildford, Surrey GU2 7XH, UK
{L.Molina-Tanco, A.Hilton}@eim.surrey.ac.uk

Abstract

In this paper we present a system that can synthesise novel motion sequences from a database of motion capture examples. This is achieved through learning a statistical model from the captured data which enables realistic synthesis of new movements by sampling the original captured sequences. New movements are synthesised by specifying the start and end keyframes. The statistical model identifies segments of the original motion capture data to generate novel motion sequences between the keyframes. The advantage of this approach is that it combines the flexibility of keyframe animation with the realism of motion capture data.

1. Introduction

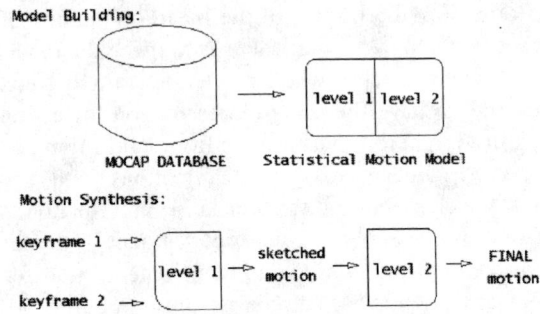

Figure 1. Overview of the system

The automatic synthesis of realistic human motion is one of the most difficult topics in computer graphics. An experienced animator is challenged when trying to create some apparently simple motion like walking. The production of only a few seconds of animation can take weeks, and re-

quires a skill that can take years to acquire. Motion capture (*Mocap*) seems to be the most suitable solution to bring realistic, natural motion into the computer when a skilled animator is not available. However, motion capture has an important weakness: it lacks flexibility. It is very difficult to edit the captured motion without degrading its quality. Commercially available motion capture libraries are difficult to use as they can have hundreds of examples which are normally browsed by the name of the actions they contain. In this paper we present a system that combines the advantages of motion capture data with those of keyframing, which is a natural and widely used interface for computer animation.

We apply statistical learning techniques to build a synthetic model from a database of motion capture examples. The model is built with two levels. In the first level, a Markov chain of the joint trajectories is built. This level on its own can generate a coarse motion by traversing states of the Markov chain. However this does not suffice if the goal is high quality, realistic animation, as all the nuances in the motion are lost in the compression performed by the statistical model. The main novelty of our approach is the second level of the model, which relates the states of the Markov chain with segments of the original motions in the database, and generates a realistic synthetic motion based on these segments.

In section 2 we compare our approach with the work of other researchers. In section 3 we explain how to build the two levels of the model. Finally we present our results and discuss the future work in sections 4 and 5.

2. Related Work

Our method derives from those employed by researchers that use state-space models to "see" human activities, following the classification provided by Aggarwal and Cai [1]. Johnson and Hogg [5] use vector quantisation and

Markov chains to build a statistical model of the silhouettes of two persons shaking hands in order to synthesise one of the silhouettes, resulting in a virtual interaction. Brand [3] builds a HMM of three-dimensional positions and velocities of motion capture markers on a human model. This model is used to infer 3D information from noisy 2D silhouettes. Both are tracking methods and therefore need a cue signal which is not available for animation. These models are not built on the joint space but on some work space of 2D or 3D points. This requires strong efforts in order to make the models invariant with respect point of view or scale.

More recently, in [4], Galata extends [5] using Variable Length HMM to model longer memory dependencies in time sequences. Also Bowden [2] uses K-means clustering and hierarchical PCA to build a Markov chain that models a motion example. These methods of motion synthesis would not be useful for animation as once a first state (keyframe) is chosen, the action that results is either fixed or random, but cannot be controlled nor can have new combinations of actions.

Pullen and Bregler [9] also use statistical models of the motion capture data, in particular a Kernel based modelling of the joint probability of features of the motion at different frequencies. They show its application to the generation of physically plausible random variations of a 2D character with 3 degrees-of-freedom performing one action. It is difficult to predict how well this method will scale for full body 3D motion.

Stuart and Bradley describe a corpus-based motion interpolation in [12]. The motion of each joint is independently learnt and represented. This allows for innovation in the synthesis, but fails to capture the correlations between joints, and very awkward poses can result.

Most of previous approaches have in common that the synthesis is an intermediate step for tracking or gesture recognition and is not a goal in itself. Hence the quality of the motion is often too poor to be of interest for character animation. However the statistical framework gives support for generalisation and handling of high dimensionality. In this paper we introduce a statistical framework which addresses the problem of "realistic" synthesis of novel human movements from a set of capture data examples.

3. Model Building

The statistical model has two levels, each having responsibility for a stage of the motion synthesis. The first level produces a sketch of the motion synthesis ("keyframes" the animation) and the second uses this sketch to sample the original motion capture data and generate a novel motion sequence.

3.1. Representation of Motion Capture Data

The raw motion capture data provides measurements of joints angles at discrete time intervals $t_i = i * \Delta t$ for $i = 0....N - 1$. A minimal **angle-axis** representation [7] is used to represent the rotation for the jth joint as a three-component vector $r_i^{\ j} = \theta(t_i) n(t_i)$ where n is a unit length axis and θ the angle about the axis. This representation gives a minimal parameterisation which readily allows the distance between two rotations and the mean of a set of rotations to be computed. For similar rotations the distance between two rotations r_a and r_b is approximated by the Euclidean distance between the vectors: $d_{ab} \approx ||(r_a - r_b)||$. The mean rotation of a set of N similar rotations can be then estimated as the barycentre $r_m \approx \frac{1}{N} \sum_{i=1}^{N} r_i$.

These properties of the angle-axis rotation enable construction of an efficient statistical representation of the motion capture data. Alternative minimal representation such as Euler angles do not allow mean rotation or distance between rotations to be directly computed. Equivalent redundant representation such as 3×3 rotation matrix or 4-component quarternion representations could be used but require additional constraints to be imposed to compute distances and to interpolate between rotations. The captured joint-angle motion data at each time instant can therefore be represented as a vector $\phi(t)$ which concatenates the individual joint angle rotations: $\phi(t_i) = [r_i^0, r_i^1, ..., r_i^{n-1}]^T$ where n is the number of joints. Translation of the root segment is ignored for the statistical model.

3.2. Preprocessing

The preprocessing comprises alignment and reduction of dimensionality of the data. Each of the motion examples is transformed by the inverse of the mean orientation of the root segment along the vertical axes to make the model **invariant** with respect to where the action is **facing** (e.g. we would like to have the same representation for a walking action from right to left as for one from left to right, where "left" and "right" are two opposite directions on the ground plane). We can represent the translation and rotation that a motion example applies to the root segment in the hierarchy of the human model via an homogeneous transformation matrix H_i^0 for each frame i. Before alignment, this transformation can be decomposed as follows[11].

$$H_i^0 = H(s_i) \, H(R_y(\theta_{3i})) \, H(R_x(\theta_{2i})) \, H(R_z(\theta_{1i})) \quad (1)$$

s_i is the translation applied to the model at frame i and $\theta_{1,2,3 \ i}$ are the Euler Angles around axes Z, X, Y. Suppose Y is the vertical axes. The mean orientation around this vertical axis can be defined as $\widehat{\theta_3} = \frac{1}{N} \sum_{i=1}^{N} \theta_{3i}$. This mean is computed for each motion example. The new transformation applied to the root segment becomes

$$H_i^{0'} = H(R_y(-\widehat{\theta}_3)) \, H_i^0 \qquad (2)$$

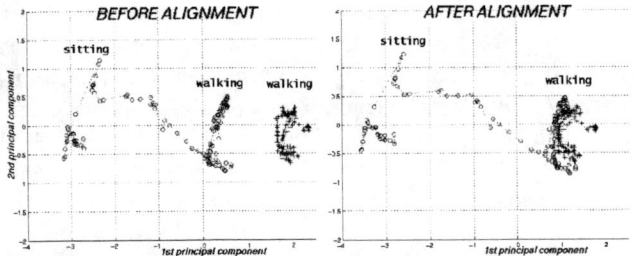

Figure 2. The two principal components of two motion examples are shown before and after alignment.

The human models used in motion capture to represent complex movements have a large number of degrees-of-freedom (DOF) (on the order of 70) corresponding to joint rotations. This representation contains considerable redundancy where joints such as hinge joints like the knee and elbow with one or two DOF are modelled as a revolute joint with 3 DOF. Also in a typical movement sequence not all physically plausible DOF are used. Motion data from optical or electromagnetic capture devices will also contain noise with a variance in the individual joint angle estimates of a few degrees. Principal Component Analysis (PCA) (see for instance [8]) is employed as a data compression technique to identify the significant variations in the data and eliminate the redundancy in the representation. The aim is to approximate N vectors ϕ in a d-dimensional space by vectors in a M-dimensional space where $M < d$. Through principal component analysis we minimise the error introduced by the dimensionality reduction by choosing the vectors to be approximated with the expression

$$\tilde{\phi} = \bar{\phi} + \sum_{i=1}^{M} a_i \mathbf{u}_i \quad with \quad \bar{\phi} = \frac{1}{N} \sum_{i=1}^{N} \phi \qquad (3)$$

where the $\mathbf{u}_i$ are the eigenvectors of the covariance matrix of the data $\mathbf{C}$. $\mathbf{C}\mathbf{u}_i = \lambda_i \mathbf{u}_i$ The eigenvectors are arranged in descending order according to the eigenvalues. The magnitude of the eigenvalues λ_i identify the components in the data which are most significant. Taking the M eigenvectors with the largest eigenvalues we can approximate the input data with a reduced basis. In practice taking approximately 15 eigenvalues represents $> 97\%$ of the variation in the mocap data. This reduces the complexity of deriving a statistical motion model from trajectories in a 70-dimensional space to a 15-dimensional space.

3.3. Level 1

After the preprocessing stage, the first level of the model building starts by dividing the joint space in clusters using a K-means classifier [6]. This is effectively a **vector quantisation** which is preferred to a multidimensional grid that divides the space into regular cells; through vector quantisation the shape of the cells can vary to adapt the density of the point distribution of the input signal. This allows us to better minimise the quantisation error [6]. A **Markov chain** is then built to recover temporal behaviour in the examples. Each state corresponds to one of the regions found in the previously computed clustering. The transition probabilities between these states are estimated by frequency counting on the training data. That is, for two consecutive input vectors found in the training data, ϕ_a and ϕ_b, first the quantised versions are computed using the K-means classifier $q_a = q(\phi_a)$ and $q_b = q(\phi_b)$. The total number of times that a transition between q_a and q_b occurs, divided by the total number of times a position is quantised as q_a, gives the estimated transition probability $P(q_b|q_a)$.

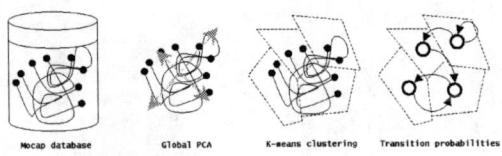

Figure 3. The first level of the model

3.4. Level 2

The second level of the model is a discrete output **Hidden Markov Model** [10] in which the hidden "states" are the motion examples. We choose the observable output probability distribution b of each of these states as being on the labels of the clusters found on the previous stage of the model, i.e. the states of the Markov chain ($b = b(cluster|example)$). This can be estimated using Bayes' relationship

$$b(cluster|example) \propto b(example|cluster)b(cluster)$$

We set the same a-priori probability for all K clusters $b(cluster) = \frac{1}{K}$. The conditional probability $b(example|cluster)$ is estimated as the ratio between the number of samples of that example that fall into that cluster divided by the total number of samples in the cluster. The result is normalised by the marginal probability

$$b(example) = \sum_{cluster=1}^{K} b(example|cluster)b(cluster)$$

139

The only thing that is needed to have a complete HMM is the transition probabilities between motion segments. These should be inversely related to the cost of jumping from one motion segment to another. In a first approximation we choose these transition probabilities to be fixed for any two examples m and n. Our basic assumption is that the longer the synthetic motion resembles a motion capture example, the more realistic it will appear. This implies choosing the transition probabilities so that we discourage jumping between motion examples, i.e. choosing $P(m|n)$ small and $P(m|m)$ large.

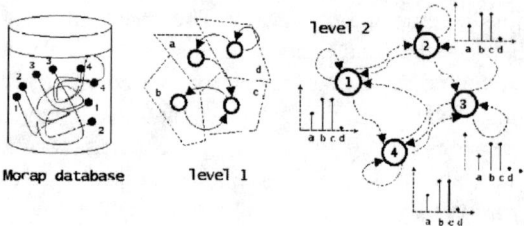

Figure 4. The second level of the model identifies motion examples with hidden states of a HMM.

3.5. Motion Synthesis

The synthesis is performed in two stages. The first stage starts by quantising the two user-specified keyframes using the K-means classifier. This results in an initial and final state in the associated Markov chain. A sequence of jumps between different states is needed to travel from the initial to the final state. We formulate this state sequence search as a **synchronous sequential decision** problem [10], which is solved using dynamic programming. Only transitions between different states have to be considered to find the most likely state sequence given the model.

A motion example may intersect several of the regions in the clustered joint space. The frontiers between these regions divide the motion examples into motion "segments". The second stage of the synthesis takes the generated state sequence as an input and solves for the most likely sequence of motion segments that could have generated that state sequence. The parameters of the HMM were chosen in such a way that the solution can be evaluated using the **Viterbi** algorithm [10] with known initial and final hidden states. The initial and final motion segments are chosen to be the motion trajectories closest to the arbitrary start and end keyframes specified by the user.

The resulting motion consists of a series of segments of the original captured motion sequences. To smoothly link these segments, the last position of the root segment in each motion segment serves as a world coordinate system for the

following motion segment. The orientations of all joints are **blended** over a small interval of frames around the point at which the two motion segments are closest. Two rotations r_a and r_b can be interpolated by taking the rotation that transforms one into the other $r_{ab} = r_b \circ r_a^{-1} = \theta_{ab}n_{ab}$. $\circ$ is the composition of rotations and r_b^{-1} is the inverse rotation of r_b [7]. Interpolated rotations b_{ab} can be found by composing a fraction $\lambda \in [0,1]$ of this rotation r_{ab} with r_a ($b_{ab} = \lambda r_{ab} \circ r_a$). The motions of a joint in two examples can be blended over an interval of time $[n, n+T-1]$ by calculating a sequence of T interpolated rotations with some function $\lambda(t_i)$ defined in that interval.

$$b_{ab}(t_i) = \begin{cases} r_a(t_i) & i \in [0, ..., n-T-1] \\ \lambda(t_i)r_{ab}(t_i) \circ r_a(t_i) & i \in [n-T, ..., n] \\ r_b(t_i) & i \in [n+1, ..., N-1] \end{cases}$$

n and $N-n+T$ are the number of frames of examples a and b respectively.

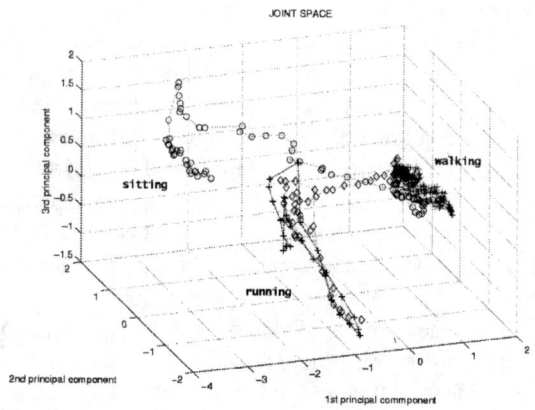

Figure 5. Three principal components of the motion capture database joint space.

4. Results

We show here an example of the functionality of the model on a small motion capture database of four motion sequences. The database contains a total of 465 frames and three different actions: standing up from the floor, walking and running. The original examples combine these actions as follows (see Fig. 5). 1: Standing up and then walking. 2: Running. 3: Walking and then running. 4: Walking.

We choose the number of clusters for the model to be 15, which represents a good compromise between accumulated quantisation error and complexity of the model. In figure 6 the error is plotted in 110 runs of the K-means algorithm as a function of K. The random initialisation of the K-means algorithm can produce big errors when K is small.

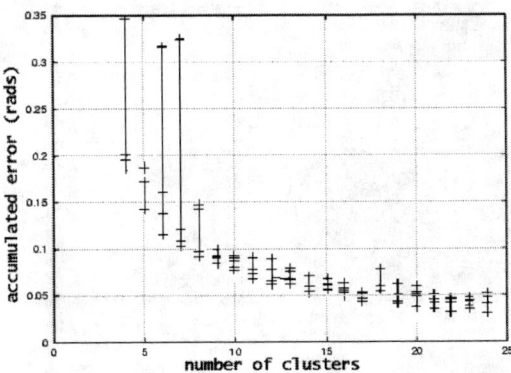

Figure 6. Accumulated quantisation errors.

A possibly desired synthetic motion would have the model standing up and then running, however his series of actions is not available in the captured examples. To synthesise the novel motion sequence we present the model with the keyframes of Fig. 7.

Figure 7. Keyframes for sitting and running.

The resulting motion has gone through a total of ten clusters and is composed of 71 frames (Fig 8). Two examples were used by the system to build the motion, number 1 and number 3. In figure 9 we show the smooth transition between the two actions.

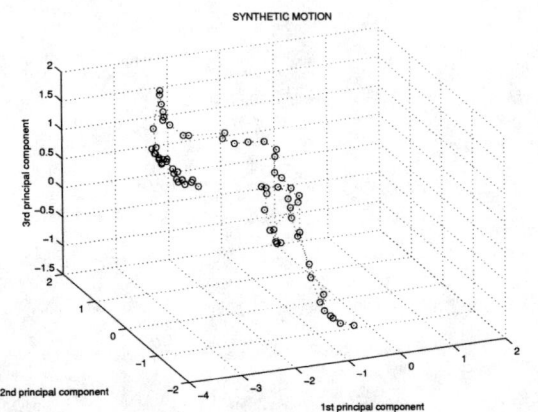

Figure 8. Synthetic motion in the joint space

5. Discussion and Future Work

We have developed a statistical framework which addresses the problem of "realistic" synthesis of novel human movements form a set of capture data examples. Our goal is to use this framework to build a natural interface between an animator and a motion capture database. For this reason we have chosen to have the system driven by user-specified keyframes.

The initial result presented in this paper for a small database of motion capture examples demonstrate the feasibility of realistic motion synthesis using a statistical model which generates novel sequences based on the original motion capture data. Results show that for this scale of database the reconstructed movement is realistic. The next stage is to scale the framework to a database of 50-100 movement examples to evaluate the performance.

Acknowledgements

The authors would like to thank Joel Mitchelson and Charnwood Dynamics for kindly providing the motion capture data. Also they would like to thank Andrew Stoddart, who initially supervised this project. This work has been funded by EPSRC grant GR/L78772 "Synthesising Human Motion for Virtual Performance".

References

[1] J. K. Aggarwal and Q. Cai. Human motion analysis: A review. *Computer Vision and Image Understanding*, 73(3):428–440, Mar. 1999.

[2] R. Bowden. Learning statistical models of human motion. In *Proceedings of the IEEE Workshop on Human Modeling, Analysis and Synthesis*, pages 10–17, June 2000.

[3] M. Brand. Shadow puppetry. In *ICCV'99. Proceedings of the International Conference on Computer Vision*, 1999.

[4] A. Galata, N. Johnson, and D. Hogg. Learning structured behaviour models using variable length markov models. In *Proceedings of the IEEE International Workshop on Modelling People (mPeople)*.

[5] N. Johnson, A. Galata, and D. Hogg. The acquisition and use of interaction behaviour models. In *IEEE Conference on Computer Vision and Pattern Recognition*, pages 866–871, June 1998.

[6] J. Makhoul, S. Roucos, and H. Gish. Vector quantization in speech coding. *Proceedings of the IEEE*, 73(11):1551–1588, Nov. 1985.

[7] X. Pennec and J. Thirion. A framework for uncertainty and validation of 3D registration methods based on points and frames. *Int. Journal of Computer Vision*, 25(3):203–229, 1997.

[8] M. Petrou and P. Bosdogianni. *Image Processing. The Fundamentals*.

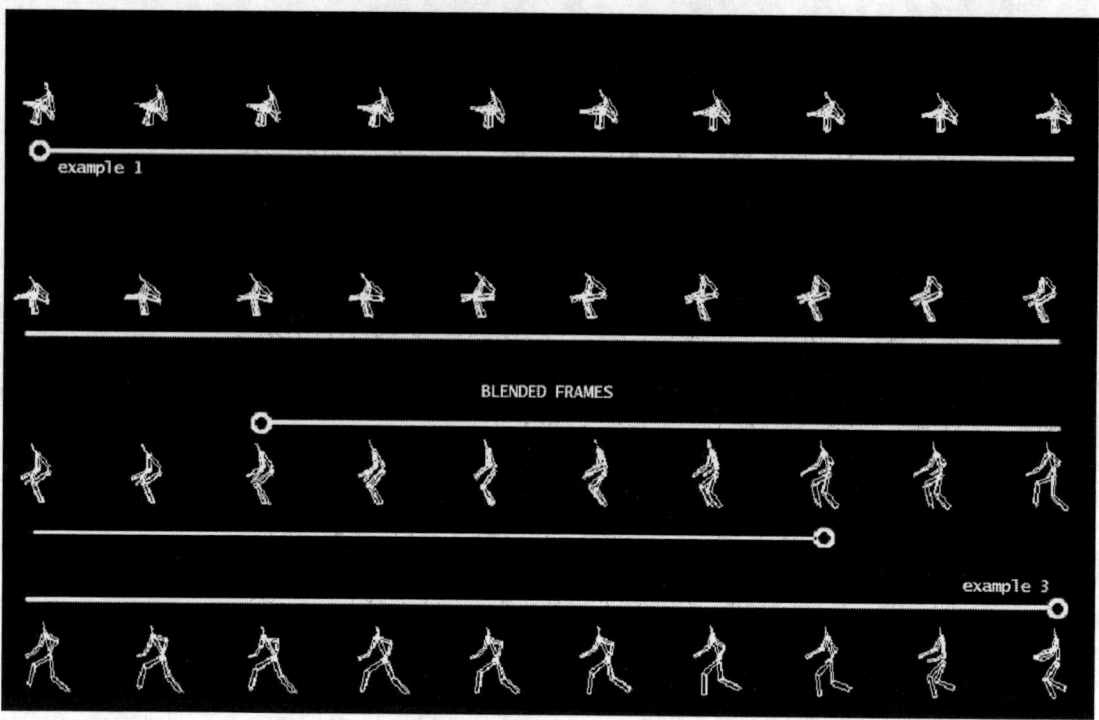

Figure 9. Transition between sitting and running

[9] K. Pullen and C. Bregler. Animating by multilevel sampling. In *IEEE Computer Animation Conference 2000*, pages 40–48, 2000.

[10] L. Rabiner and B.-H. Juang. *Fundamentals of Speech Recognition*. Prentice-Hall, 1993.

[11] K. Shoemake. *Graphics Gems IV*, chapter Euler Angle Conversion, pages 222–229. Academic Press, 1994.

[12] J. M. Stuart and E. Bradley. Learning the grammar of dance. In *Proc. 15th International Conf. on Machine Learning*, pages 547–555. Morgan Kaufmann, San Francisco, CA, 1998.

Face Detection and Attentional Frames for Visually Mediated Interaction

A. Jonathan Howell and Hilary Buxton

School of Cognitive and Computing Sciences,
University of Sussex, Falmer, Brighton BN1 9QH, UK

{jonh,hilaryb}@cogs.susx.ac.uk

Abstract

In this paper we introduce a set of effective computational techniques for understanding the visual aspects of human interaction which could be used, for example, in video-conferencing applications. First, we present methods for face detection and capture of attentional frames to focus the processing for visually mediated interaction. Second, we present methods for recognising the various gesture phases that can be used to control the camera systems in the integrated system. Finally, we discuss how these techniques can be extended to 'virtual groups' of multiple people interacting at multiple sites.

1 Introduction

Many real-world applications of computer vision require understanding of human motion and behaviour. In particular, using visual interfaces in *Visually Mediated Interaction* (VMI) is a key area of research for applications in video telecommunication. These typically exploit face and gesture recognition techniques with some kind of intelligent motion-based control of attention to track interacting participants in the dynamic scenes.

In general, robust tracking of non-rigid objects such as human bodies is nontrivial due to rapid motion, occlusion and ambiguities in segmentation and matching. On-going research at the MIT Media Lab has shown progress in the modelling and interpretation of human body activity [19, 24] especially when using computationally simple view-based approaches to action recognition [1]. Similar attempts at Microsoft Research [3, 22] have also yielded useful results. Our aim is to go further into the task of *intentional control* in visual communication using head-pose together with a set of 'interaction relevant' gestures.

First, however, this paper concentrates on the real-time extraction of an attentional frame that tracks the head and hands of participants in the visual interaction. In section 2 we introduce the RBF and TDRBF networks [8, 9, 16, 17] and training databases used in these studies. Then in sec-

tion 3 we describe the real-time capture of the attentional frame using motion and skin colour features to define a window that can be normalised using either resampling or active camera techniques. This leads into the computationally cheap RBF-based face detection and pose estimation process discussed in section 4.

Second, for behavioural modelling to drive real-time visual interaction, we need to specialise the pre-, mid- and post- phases of the pointing and waving gestures we want to use in our visual interactions. This will allow us to initiate and terminate control of system activity such as changing the direction and scene focus of our cameras. The VMI framework we have been developing [21] requires high-level information about group and individual interaction in a 'scene vector' to provide a 'camera control vector'. The scene vector typically provides ongoing probabilities of the dynamic head-pose and gesture phases for interacting participants and the camera control vector typically provides reactive direction and zoom. Here a TDRBF-based scheme for extracting the scene vector and predictive control using an adaptive TDRBF mapping onto the camera control vector are discussed in section 5 on the integrated system. Section 6 summarises some of the most important observations from these studies and section 7 concludes with a discussion of our ongoing research plans and general issues that need to be addressed in the extension of these techniques to 'virtual groups' of multiple people interacting at multiple sites.

2 The RBF and Time-Delay RBF Model

The RBF network is a two-layer, hybrid learning network [16, 17], which combines a supervised layer from the hidden to the output units with an unsupervised layer from the input to the hidden units. The network model is characterised by individual radial Gaussian functions for each hidden unit, which simulate the effect of overlapping and locally tuned receptive fields.

To construct a dynamic neural network, recurrent connections can be added to standard multi-layer perceptrons which then form a contextual memory for prediction over time [4, 12, 18]. These partially recurrent neural networks can be trained using back-propagation but there may be problems with stability and very long training sequences when using dynamic representations. Instead, we use a simple Time-Delay mechanism in conjunction with an RBF network, which we term a TDRBF network, to allow fast, robust solutions to difficult real-life problems. The Time-Delay Neural Network (TDNN) model (for an introduction, see Hertz et al. [5]), incorporates the concept of time-delays in order to process temporal context, and has been successfully applied to speech and handwriting recognition tasks [23]. Its structured design allows it to specialise on spatio-temporal tasks, but, as in weight-sharing network, the reduction of trainable parameters can increase generalisation [13].

A Time-Delay RBF network can be created by combining data from a fixed time 'window' into a single vector as input. Our approach, successful in previous work with RBF networks for face recognition tasks with image sequences [8, 11], uses an RBF unit for each training example, and a simple pseudo-inverse process to calculate weights.

2.1 The Standard Gesture Database

For our gesture recognition experiments, we used the IS-CANIT Phase I gesture database [10], which was created in the ISCANIT project with Shaogang Gong and Jamie Sherrah at Queen Mary and Westfield College, London, who are researching real-time face detection and tracking [15, 20]. To use this as a standard database for this work allows us to present comparative results.

The database contains four examples of four gestures from three people, 48 5-second sequences in all. The gestures are *pntrl*, pointing left, *pntrr*, pointing right, *wavea*, waving hand above head, and *waveb*, below head. For the work presented here, we use one wave (*waveb*) only, as there was some doubt about the semantic usefulness of two waves plus difficulty in determining how to exactly define and, therefore, distinguish them. In addition, we have altered the training of the gesture information to reduce the amount of data required. To do this, we took advantage of the tri-phasic nature of the waving and pointing gestures to split each gesture into a pre-, mid- and post-gesture sequence, each of which can be used for training as a separate sub-gesture class.

When considering the particular tri-phasic gestures in this database, it may be that the post-gesture data does not contribute useful information to the recognition system, for example, the end of a wave gesture is when the side-to-side hand movement stops, not when the hand returns to the person's lap. For this reason, we have tested the system in two configurations: training with all three phases, and training with just the first two (pre- and mid-gesture), as a further reduction in training data. All sub-gesture sequences were

Figure 1. Use of colour/motion information to position an attentional frame around a person: (a) a box is centred around each colour/motion 'blob', the inner vertical lines representing the standard deviation of the pixels along the x-axis, giving a width measure, (b) having identified which box contains the head (the uppermost one in (a)), an attentional frame box is drawn around the person relative to the head position, and sized according to head width. The top right image shows the image area inside the head box, bottom right the resampled area of the image inside the attentional frame.

fixed in length at 7 frames, which was the shortest phase found in the training database, as a tradeoff between efficiency, in having shorter sequences (taking up less memory), and completeness, where ambiguity may arise through overlapping parts of other phases being included in training data.

This gesture-phase training gives some predictive power from the pre-gesture and focusses on the characteristic movement of the mid-gesture. It also gives some temporal invariance through allowing the two phases of the gesture to be different in length to each other (the original studies imposed a fixed relationship between the overall gesture length and its three phases). By not waiting for the post-gesture phase, we obtain more rapid signalling of the gesture event.

An integration layer on the TDRBF network can be used to combine results from successive time windows, which will give smooth gradations between serial actions. Here we know each sequence contains only one action, and so can rely on our temporal segmentation to give the single best frame position for classification. A sparse arrangement of Gabor filters [6] is used to preprocess each frame of the sequences (colour/motion information for gestures, grey-level pixels for face detection): data is sampled at three non-overlapping scales and three orientations with sine and cosine components for a total of 126 coefficients per frame.

3 Capturing the Attentional Frame

We will use colour/motion cues from the image sequence to identify and track the head. Once we know the position

Search	Tracking Position % Correct
Global Search	78
Local Search	100
Global/Local Search	86
Temporal Matching	97

Table 1. Accuracy in head tracking using colour/motion blobs: Global search checks all blobs every frame, Local search looks only for blobs within a few pixels of the specific head blob in the previous frame, Global/Local search initially uses a global search, but will switch to local search when possible, Temporal Matching is a global search, but tracks only blobs matched to nearby blobs from previous frames.

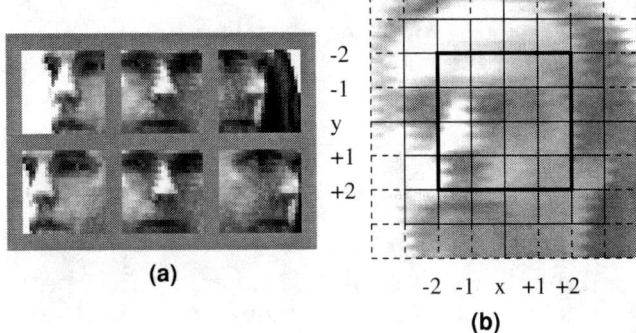

(a)

(b)

Figure 2. (a) Two methods for segmenting 25×25 pose-varying face data: (top row) nose-centred, (bottom row) face-centred, the former being used for experiments here (b) the grid system for detecting potential faces within a potential 'head blob' region of the image: each area tested is represented by a 4×4 box, the thick line shows the central position ($x, y = 0$), normal line and dashed lines the outer positions 1 and 2 spaces out from the centre. In this case, a maximum output would be expected at $x = -1, y = +1$, which indicates a head-pose slightly down and turned to the right.

and size of the head, we can define an 'attentional frame' region around the person. The attentional frame is a 2-D area around that person that contains all the body movement information relevant to our application, which is all movement of the head and right arm. To allow people to move closer or further away from the camera, this information is normalised for size, relative to head size, around an arbitrary standard position around 3m from our Sony EVI camera at its shortest focal length (5.4mm). With this arrangement, we can segment people within 1.5–7m from the camera. This gives a resampled image size for the head of about 30 pixels across, with the face information 25 pixels across.

Our main priority is to find real-time solutions for our application. Therefore, we use two computationally cheap pixel-wise processing techniques on our image: thresholded frame differencing, giving motion information, and Gaussian mixture models [15], giving skin colour information. These are combined to give a binary map of moving skin pixels within the image, and we use local histogram maxima to identify potential 'blob' regions. A box which is large enough to contain the head at all distances in our target range is then fitted over each of these regions. Fig. 1(a) shows how each box is centred on the centroid of each maximum, with the inner lines showing the standard deviation of the pixels along the x-axis from that centroid. It can also be seen that the hands are ignored in this example, as they are too low down to be included in a face-size 'blob'.

If we assume these blobs will have a roughly Gaussian profile, the pixels within the blobs that represent actual parts of people within our usable distance range (defined above) will have a standard deviation that is also within a fixed range. In Fig. 1(a), there are four potential blobs, two of which have been discarded (represented by being drawn with a dashed line) for having a width standard deviation that is too high. The width of the blob is a more reliable

measure of the head size than height, as the latter may be altered by clothes (varying the amount of skin shown around the neck) and hairstyle (varying the amount of skin seen on the forehead). The position of the head is initially set to the blob with the greatest area, as the head is generally larger than the hands. This can be verified subsequently in the face detection stage.

Table 1 shows that a reasonable level of performance can be obtained with a global search of the image each frame. If the initial position of the head is provided, a local search was very much better, being able to cope with quite large changes in position and size. Such approaches are too brittle for real-life applications, however, because if the head is ever lost (becomes occluded or goes off screen), there is very little chance of reestablishing its position afterwards. A compromise is to use both methods, checking potential head blobs over the whole image, but limiting large changes in position until the competing candidate becomes more stable over time than the current most-likely head blob. A more robust approach was what we termed 'Temporal Matching': here the tracker will only consider blobs from the current frame which have been matched to nearby blobs from previous frames. This excludes any anomalous blobs that appear for one frame only in an image sequence, and promotes those that exhibit the greatest temporal coherence.

Having found the position and size of the head, we can

			Face Detection			
Sampling	Non-Face Blobs	Face/ Non-Face	% Correct Pre-Discard	% Dis-carded	% Correct Post-Discard	Pose % Correct
3×3	1	180/180	41	18	50	73
	2	180/360	68	23	88	73
5×5	None	180/320	6	0	6	74
	1	180/820	90	5	95	80
	2	180/1320	91	5	95	82

Table 2. Results for face detection RBF networks with different face/non-face training schemes. Right-hand column indicates success of pose estimation, based on offset of nose-centred face region relative to the whole 'head blob'. These results use a global search on each frame (a local search based on temporal continuity would reduce errors).

extract the attentional frame from around the person, see Fig. 1(b). Section 5 discusses the use of this information for gesture recognition.

4 Pose-Invariant Face Detection

The previous section described how we isolate small areas of moving skin-tones from the overall image. This reduces computation and network size, by allowing the face detector to work only within a small subset of the full spectrum of possible objects typically encountered in an office environment. Specifically, we can consider the restricted form of face detection where we need to distinguish a face only from other moving skin-tone blobs (typically hands).

In order to perform face recognition later, we need to identify the position of the central face area (eyes, nose, mouth), rather than the entire skin area on the head (which also includes forehead, neck, ears, etc). Our face detection task, therefore, is to distinguish centred faces (of any head pose) from both non-centred faces and other moving skin-tone blobs only. This is in contrast to more conventional face detectors, which have to find (frontal) faces in any patch of image. We train a RBF network with examples of both to provide two competing 'face/non-face' outputs, with a level of confidence based on the difference between the two output values from the network [7]. This level of confidence allows discarding of low-confidence results where data is noisy or ambiguous.

Our training examples need to take variable head-pose into account, so the central face region of a person can be recognised at all normal physiological pose positions. Facial information is only visible on a human head from (roughly) the front ±120° of x- and y-axis movement, and z-axis movement is physiologically constrained to around ±20° (when standing or sitting) [6]. The face region is centralised on the nose, rather than the face, for all profiles, as this allows non-occluded face information to remain roughly in the same position, see Fig. 2(a). This has previously been shown to more useful for pose-varying face

recognition [6].

The face detection RBF network was trained with 20 frames from an image sequence that contained a full range of y-axis head-poses from left to right. The position of the nose was determined manually and a 3×3 grid of 25×25 face images was extracted from each frame for the 'pro-face' class. In addition, a number of non-face regions were extracted for the 'anti-face' class. These were of two types: (a) from the periphery of the face (to encourage face detection only where the image was accurately aligned on the face), and (b) from other skin/colour regions, specifically the hands. A 5×5 testing grid was found to be the most useful for our tests: 3×3 did not give sufficient movement across the head to cover the full pose range, and 7×7 gave spurious output when the network was applied to areas of the image that did not contain face or hands.

Table 2 shows that the 5×5 trained network performed well, but only if extra non-face examples from the hands were used for training. The 3×3 network, which had no mal-aligned examples of faces, did not perform as well, but was able to discard results through low-confidence to give better performance.

We can determine a coarse estimate of head-pose, such as left, frontal or right, from the output grid (see Fig. 2(b)). This qualitative level of head-pose is very useful for group interaction analysis [21].

The conclusions we draw from these results are that the use of hand images for the 'anti-face' class allowed the network to train much more efficiently than when using mal-aligned face images alone. In addition, the performance of the more computationally efficient 3×3 trained network may be sufficient for real-time use where we can take advantage of temporal continuity to constrain the search area for heads.

5 The Integrated System

A complete video-conferencing active camera control system requires high-level interpretation of group and in-

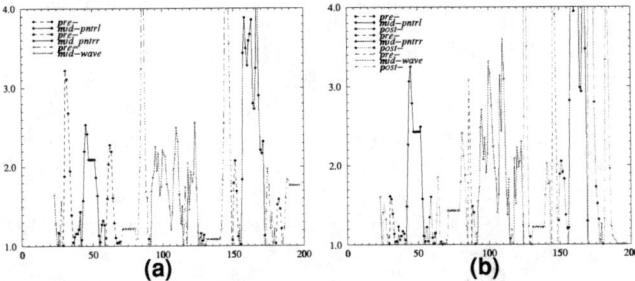

Figure 3. Output for the TDRBF gesture network with a free gesture test sequence with different background, lighting and person to that encountered during training. Frame numbers for the start of the three phases (plus overall end) of each gesture: *pntrl* **24, 34, 54, 64,** *wave* **80, 90, 117, 124,** *pntrr* **140, 155, 170, 183. Each line shows the confidence level of output when its class is the maximum, and is zero at all other times: (a) Two-phase training: each line represents a pre- or mid-gesture class, (b) Three-phase training: each line represents a pre-, mid- or post-gesture class.**

dividual interaction [21]. We have to provide sufficient information, contained in a 'scene vector' [21], for this interpretation to take place and for the system to provide camera control information, the 'camera control vector'. The 'scene vector' typically provides head-pose and gesture probabilities for the people in the field of view. If individuated control of the system is required, then we need to identify who these people are (from a small known group). Two extra stages, therefore, are needed: gesture and (pose invariant) identity recognition. We have previously shown practical techniques for tackling these tasks in real-time, using the RBF and TDRBF networks [8, 10].

To test our gesture recognition network, we used a sequence originally used to test algorithms using the Phase I data [14], which contains 197 frames with different background, lighting and person to that in the database.

Fig. 3(a) shows how a trained TDRBF network works with these smaller, phase-based classes. One aspect we had not anticipated is that the post-gesture phase, which is not explicitly trained for, is often picked up by the network as a pre-phase. To try to reduce this effect, we also trained the network with all three phases of each network (producing nine classes, and the output from this network is shown in Fig. 3(b). The mid-classes are unambiguously recognised in both types of training, which is the purpose of the network. Some false pre-gesture 'hypotheses' are raised by the network, but these are replaced by the correct pre-gesture

classification before the mid-gesture occurs. It might be expected that the pre-gesture classes will be confused, especially when they first start, as raising your hand to wave or point will initially both look very similar, but the results here indicate that they are correctly classified in time for predictive control.

To complete our integrated system, we need to pass this gesture and head-pose information, with identity if appropriate, to a higher-level interpretation network [21]. In addition, we will have to adapt our system to cope with multiple people in the scene, which increases the complexity of the low-level processing stage. There will be more head blobs to find, but by assigning attentional frames to each person, and analysing each of these separately, it is hoped that problems due to occlusion from other members of the group will be kept to a minimum.

6 Observations

Several points can be seen from the results:

- We can use colour/motion cues to effectively segment and track human heads in image sequences.
- An attentional frame can be extracted relative to the head position and size to allow the real-time recognition of hand gestures through time.
- By extracting colour/motion regions from the overall image, the face detection task is greatly simplified.
- A face detection network can be used to give a qualitative estimate of head-pose for predictive control.
- Splitting multi-phasic gestures into separate phase classes not only gives more precise timing of gesture events, but also allows the gesture recognition network to provide prediction hypotheses by identifying pre-gesture classes.

7 Conclusion

In previous papers, we have shown how single and multi-person activity can be learnt from training examples and generalised to different instances of the same kind of activity [10, 21]. Here we have shown how the approach can be scaled up to more general application and work in real-time by specialising recognition of different phases of the gesture. In future work we hope to show extrapolation to novel scenarios in addition to the generalisation to different people and backgrounds demonstrated in this paper. We have also demonstrated how our connectionist techniques can support an integrated real-time system by detecting faces and capturing the associated attentional frames to focus the processing for visually mediated interaction (VMI).

A significant issue to be addressed in future work is whether these adaptive learning techniques have sufficient power to overcome the ambiguities present when multi-person activity takes place. Currently, the system relies on several independent components so the probability of failure of one of these is quite possible. The gesture-based interaction must, therefore, be able to cope with missing or noisy inputs, for example, if a head is occluded. One way to guide the system's attention is to feed back higher-level expectations to the low-level tracking and detection process, i.e. predictive mapping of what to look for in space and time. This certainly improves intelligent control of attention in dynamic scenes [2]. In any event, we will have to design our VMI systems around the task demands which include both the limitations of our techniques and potentially conflicting intentions from the participants. The connectionist techniques proposed here are arguably best suited to this kind of situation as they can learn adaptive mappings and have inherent constraint satisfaction.

Another key issue for future work is to extend the techniques to deal with 'virtual groups' of people communicating at different sites. In our proposed future work, we will be developing and evaluating real-time user behaviour models based on temporal prediction of continuous pose and gesture change [21] to cope with such virtual groups. We have been able to develop appropriate connectionist techniques for all levels of the system from face detection and tracking with feedback control, through to recognition, if appropriate, of faces [9], poses [6], expressions [7] and gestures [10], plus the active *intentional control* of attention by the users demonstrated here. To cope with virtual groups, we will require more powerful prediction capabilities as well as fundamental extensions of the behaviour modelling and interactive control of attention. The system will remain essentially multiple view-based in order to provide real-time behaviour interpretation and prediction. Our approach will also continue to use task-specific representations and algorithms in order to avoid unnecessary computational cost in dynamic scene interpretation [2].

Acknowledgements

The authors gratefully acknowledge the invaluable discussion, help and facilities provided by Shaogang Gong, Stephen McKenna and Jamie Sherrah during the development and construction of the gesture databases.

References

[1] A. F. Bobick. Movement, activity, and action: The role of knowledge in the perception of motion. *Proc. Royal Society London, Series B*, 352:1257–1265, 1997.

[2] H. Buxton and S. Gong. Visual surveillance in a dynamic and uncertain world. *Artificial Intelligence*, 78:431–459, 1995.

[3] R. Cutler and M. Turk. View-based interpretation of real-time optical flow for gesture recognition. In *FG'98*, pp. 416–421, Nara, Japan, 1998.

[4] J. Elman. Finding structure in time. *Cognitive Science*, 14:179–211, 1990.

[5] J. A. Hertz, A. Krogh, and R. G. Palmer. *Introduction to the Theory of Neural Computation*. Addison-Wesley, CA, 1991.

[6] A. J. Howell. *Automatic face recognition using radial basis function networks*. PhD thesis, University of Sussex, 1997.

[7] A. J. Howell. Face recognition using RBF networks. In R. J. Howlett and L. C. Jain, editors, *Radial Basis Function Networks: Design and Applications*. Springer-Verlag, 2000 (In Press).

[8] A. J. Howell and H. Buxton. Towards unconstrained face recognition from image sequences. In *FG'96*, pp. 224–229, Killington, VT, 1996.

[9] A. J. Howell and H. Buxton. Recognising simple behaviours using time-delay RBF networks. *Neural Processing Letters*, 5:97–104, 1997.

[10] A. J. Howell and H. Buxton. Learning gestures for visually mediated interaction. In *BMVC'98*, pp. 508–517, Southampton, UK, 1998.

[11] A. J. Howell and H. Buxton. Learning identity with radial basis function networks. *Neurocomputing*, 20:15–34, 1998.

[12] M. I. Jordan. Serial order: A parallel, distributed processing approach. In *Advances in Connectionist Theory: Speech*. Lawrence Erlbaum, NJ, 1989.

[13] Y. Le Cun, B. Boser, J. S. Denker, D. Henderson, R. E. Howard, W. Hubbard, and L. D. Jackel. Backpropagation applied to handwritten zip code recognition. *Neural Computation*, 1:541–551, 1989.

[14] S. J. McKenna and S. Gong. Gesture recognition for visually mediated interaction using probabilistic event trajectories. In *BMVC'98*, pp. 498–507, Southampton, UK, 1998.

[15] S. J. McKenna, S. Gong, and Y. Raja. Face recognition in dynamic scenes. In *BMVC'97*, pp. 140–151, Colchester, UK, 1997.

[16] J. Moody and C. Darken. Learning with localized receptive fields. In *Proc. 1988 Connectionist Models Summer School*, pp. 133–143, 1988.

[17] J. Moody and C. Darken. Fast learning in networks of locally-tuned processing units. *Neural Computation*, 1:281–294, 1989.

[18] M. C. Mozer. Neural net architectures for temporal sequence processing. In *Time Series Prediction: Predicting the Future and Understanding the Past*, pp. 243–264. Addison-Wesley, CA, 1994.

[19] A. Pentland. Smart rooms. *Scientific American*, 274(4):68–76, 1996.

[20] J. Sherrah and S. Gong. Exploiting context in gesture recognition. In *Proc. 2nd Int. Interdisciplinary Conf. Modelling and Using Context*, Trento, Italy, 1999.

[21] J. Sherrah, S. Gong, A. J. Howell, and H. Buxton. Interpretation of group behaviour in visually mediated interaction. In *ICPR'2000*, Barcelona, Spain, 2000.

[22] M. Turk. Visual interaction with lifelike characters. In *FG'96*, pp. 368–373, Killington, VT, 1996.

[23] A. Waibel, T. Hanazawa, G. Hinton, K. Shikano, and K. Lang. Phoneme recognition using time-delay neural networks. *IEEE Trans. Acoustics, Speech, & Signal Processing*, 37:328–339, 1989.

[24] C. R. Wren and A. P. Pentland. Dynamic models of human motion. In *FG'98*, pp. 22–27, Nara, Japan, 1998.

Real-time Human Motion Analysis and IK-based Human Figure Control

Satoshi Yonemoto,   Daisaku Arita   and   Rin-ichiro Taniguchi
Division of Intelligent Systems, Kyushu University
6-1 Kasuga-koen Kasuga Fukuoka, 816-8580 Japan
{yonemoto, arita, rin}@limu.is.kyushu-u.ac.jp

Abstract

This paper presents real-time human motion analysis based on real-time inverse kinematics. Our purpose is to realize a mechanism of human-machine interaction via human gestures, and, as a first step, we have developed a computer-vision-based human motion analysis system. In general, man-machine 'smart' interaction requires real-time human full-body motion capturing system without special devices or markers. However, since such vision-based human motion capturing system is essentially unstable and can only acquire partial information because of self-occlusion, we have to introduce a robust pose estimation strategy, or an appropriate human motion synthesis based on motion filtering. To solve this problem, we have developed a method based on inverse kinematics, which can estimate human postures with limited perceptual cues such as positions of a head, hands and feet. In this paper, we outline a real-time and on-line human motion capture system and demonstrate a simple interaction system based on the motion capture system.

1   Introduction

Man-machine seamless 3-D interaction is an important tool for various interactive systems such as virtual reality systems, video game consoles, etc. To realize such interaction, the system has to estimate motion parameters of human bodies in real-time. Up to the present, as a method for human motion sensing, many motion capture devices with special markers or magnetic sensor attachments have been employed. Since they need special marker-sensors, they often impose physical restrictions on the object. On the other hand, recently, fully image-feature-based motion capturing systems which do not impose such restrictions have been developed as a computer vision application[1]. Although the vision-based approach still has problems to be solved, it is a very smart approach which can achieve seamless human-machine interaction. Moreover, it has a merit that it can acquire shape properties and surface textures, which can not be measured by the former approach. Therefore, we are undertaking to develop an image-feature-based motion capturing system, giving consideration to alleviating scene constraints and physical constraints imposed on the system as little as possible.

To analyze human motion, image features such as blobs (coherent region)[1][2][3] or silhouette contours are usually employed. Since the contour-based image features essentially depend on human postures, they are appropriate only for the estimation of typical postures. Therefore, many researchers have developed skin-color region tracking and stereo reconstruction methods using general region clustering. In particular, *Pfinder*[1] has shown that blob tracking is applicable for many real-time applications although the idea is not very new. Recently, *purposeful human motion*[4] employing the above blob tracking has been proposed. It is based on an analysis and synthesis framework with fast dynamics engine[5], and with an HMM based multiple behaviour model. The method is applied to upper body motion estimation, and temporary occlusion in the intersection between blobs of both hands or head-and-hand is handled, although the method does not solve the corresponding problem. In gesture recognition systems, as well as in human motion tracking, human motion primitives are also dealt with. However, in these systems, gesture representation is symbolically defined and such symbol-based approaches are not appropriate for our purpose, which has to generate actual 3-D body postures, or 3-D positions of head, arms, feet, etc.

In this paper, we present vision-based human motion capture system based on inverse kinematics, which can estimate human postures with limited perceptual cues such as positions of a head, hands and feet. We also focus on on-line implementation of the motion capture system using a PC-cluster (multiple PCs connected via high-speed network).

2   System Overview

The flow of our algorithm of real-time motion capturing is as follows:

1. Detection of cues (*perception*)

   - 2-D color blob tracking for each view
   - Calculation of 3-D color blob position using multi-view fusion

2. Generation of human figure full-body motion and rendering in the virtual space and calculation of the interaction (*motion synthesis*)

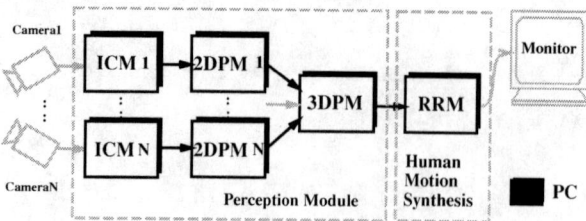

Figure 1: Image processing modules on PC cluster.

A prototypical system developed here is a real-time visually-guided-animation system, and make it real-time and on-line, ew have implemented the system on a PC cluster, which consists of multiple PCs connected via a high-speed network, *myrinet*[7]. Each pipeline step is controlled by a synchronization mechanism[8]. Fig.1 shows the system flow and allocation of processing modules to PCs.

Details of the processing modules are as follows:

a) Perception Module:

Image Capturing Module(ICM) These modules work as image-capturing and resizing modules (320 × 240). Each ICM_v ($v = 1, \cdots, V$; V is the number of cameras) sends image data to $2DPM_v$.

2-D Processing Module(2DPM) These modules work as 2-D image processing modules (2-D blob tracking). Each 2DPM receives the image data from ICM_v, and sends 2-D extracted image feature data (positions of the center of gravity of the 2-D blobs) to 3DPM.

3-D Processing Module(3DPM) This module works as a 3-D vision processing module. It receives and integrates the 2-D image feature data from $2DPM_v$ ($v = 1, \cdots, N$), and estimates 3-D model parameters (3-D positions of blobs). The estimated parameters are sent to the RRM.

b) Human Motion Synthesis:

Real-time Rendering Module(RRM) This module works as a real-time renderer of the virtual space. It receives the 3-D blob positions from 3DPM and estimates 3-D pose and motion of the human body based on the received data.

In the following sections, we will show details of the algorithms and some experimental results.

3 Perception

3.1 Color Identification

In this system, skin color regions observed in an input image are interpreted as hand and face(head) blobs, regions with pre-acquired shoe or sock color as feet, and a region with shirt color as a torso

blob(Fig.2). We assume the colors can be represented in a simple parametric form which is relatively robust for illumination changes[6]. In other words, we assume the color features (r,g,b) of each pixel are represented in the following quadratic equations of intensity i:

$$
\begin{aligned}
\hat{r} &= R_2 i^2 + R_1 i, \\
\hat{g} &= G_2 i^2 + G_1 i, \\
\hat{b} &= B_2 i^2 + B_1 i
\end{aligned}
\tag{1}
$$

For each blob color to be identified, six model parameters, or coefficients, $R_1, \cdots, B_2$ are estimated in advance from a training data set, or real blob images. In color identification, the system computes the following error between observed color features (r, g, b) of a pixel and the model color features $(\hat{r}, \hat{g}, \hat{b})$ calculated, according to the above equation, from the intensity i of the pixel.

$$
error = (r - \hat{r})^2 + (g - \hat{g})^2 + (b - \hat{b})^2
\tag{2}
$$

The system identifies the color of a pixel as a color giving the minimum error that is less than a certain threshold.

3.2 2-D Blob Tracking

Blob tracking is accomplished according to the following steps:

1. A rectangle containing a human body is detected after background image subtraction and thresholding are applied. Then, regions with skin color, shoe/sock color or shirt color are identified by the above method. At the same time, the torso position (the center of gravity) is estimated.

2. In the rectangle, the color-identified pixels are classified into blobs based on the similarity of their colors and positions to those of blobs detected in the previous frame.

Here, we only assume that the result of background image subtraction is stable, and that cloth color is not similar to skin color. These assumptions can be easily met, particularly, in indoor or studio-like situations.

Initial correspondence of color-identified regions to specific 2-D blobs, i.e., a face, hands and feet, is decided when the system starts up, based on simple heuristics of the natural standing position. The heuristics employed here are as follows:

- The head is a skin color region upper-most in the detection rectangle.

- The left(right) hand is a skin color region which is on the left(right) in the detection rectangle.

- The left(right) foot is a pre-acquired color region which is on the left(right) in the detection rectangle.

This correspondence is also examined when the system fails to track the blobs. The error recovery process is quite important for online algorithms, and the decision process should be carefully designed.

3.3 Estimation of 3-D Blob Position

When a 2-D blob is detected in two views, the 3-D position of the blob can be calculated by a stereo method. However, since self-occlusion often occurs, with only two views it is almost impossible to estimate all parts of the moving body for a long period. Therefore, multi-view fusion is indispensable. In the blob tracking, precise estimation is not required and, therefore, we have employed a simple but fast multi-view fusion strategy. The algorithm of 3-D blob position calculation adopted here is as follows:

Selection of views According to the visibility of views, reliable views, or views whose visibility is higher than a certain threshold, are selected for each blob. The visibility is defined as the number of observed pixels in each blob, and it can indicate whether occlusion is occurring or not.

Calculation of line of sight According to camera calibration information, for each of the selected views, a line of sight, or a vector from the origin of the camera coordinate system to the center of gravity of the blob, is calculated.

Integration of multi-view information
Referring to the acquired lines of sight, the 3-D position of each blob is calculated.

When a line of sight calculated for the most reliable view is parameterized as $\mathbf{T}_1 = \mathbf{o}_1 + t_1\mathbf{d}_1$ (t_1 is a parameter), and the rest of the lines of sight as $\mathbf{T}_j = \mathbf{o}_j + t_j\mathbf{d}_j$ (t_j is a parameter; $j = 2, \cdots, J$), the intersection point $\mathbf{T}$ is approximated as a point on the line of sight $\mathbf{T}_1$ whose average distance to the other lines of sight is smallest in the sense of the least squares error.

$$\mathbf{T} = \mathbf{o}_1 - \frac{\sum_{j=1}^{J}(\mathbf{d}_1 \times \mathbf{m}_j, \mathbf{o}_1 \times \mathbf{m}_j - \mathbf{n}_j)}{\sum_{j=1}^{J}||\mathbf{d}_1 \times \mathbf{m}_j||^2}\mathbf{d}_1, \quad (3)$$

where

$$\mathbf{m}_j = \frac{\mathbf{d}_j}{\sqrt{1+||\mathbf{o}_j \times \mathbf{d}_j||^2}}, \qquad \mathbf{n}_j = \frac{\mathbf{o}_j \times \mathbf{d}_j}{\sqrt{1+||\mathbf{o}_j \times \mathbf{d}_j||^2}}.$$

The calculated point $\mathbf{T}$ corresponds to the 3-D blob position $(T_x, T_y, T_z)^T$.

4 IK-based Motion Synthesis

Information acquired in the perception process is just 3-D positions of blobs, which correspond to a torso, a head, hands and feet of a human body. Therefore, to estimate the body posture from these cues, the number of which is less than the degree of freedom of the body, we have to solve the inverse kinematics[9]. In our case, a human body is represented as a multi-part articulated object, or as 14 parts with 23 degrees of freedom (see Fig.3), and the 3D blob positions are given as the goal positions, or the end effectors. Of

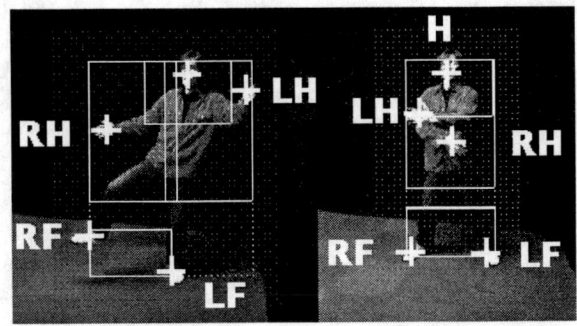

Figure 2: 2-D blob position estimation results.

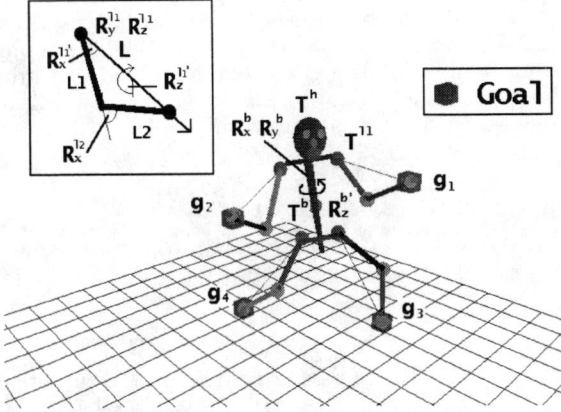

Figure 3: Our human figure model geometry.

course, there are approaches in which knees and elbows are detected based on contour analysis, or silhouette analysis, but they cannot stably detect those positions for many postures.

The goals of the inverse kinematics which we have designed can be summarized as follows:

- Inverse kinematics of four connecting links, which are two arms and two legs, can be solved in real-time.

- Even when goal positions (3-D blob positions) given by the perception module are not precise, a solution can be derived to some extent.

- The solution gives us continuous and natural-looking motion of the human body.

In our case, as mentioned above, the 3-D blob positions acquired by the perception modules are sometimes imprecise. In other words, the goal positions are sometimes established at positions where physically possible solutions cannot be derived. Therefore,

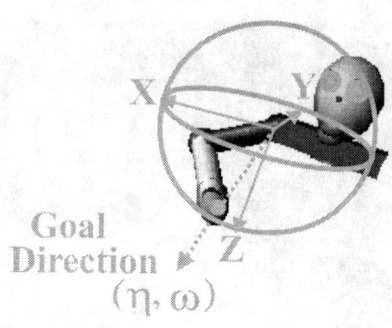

Figure 4: Definition of a goal with arm direction.

we interpret each of the given goals as the combination of the direction of the goal and the distance to the goal. When the goal position is located where a physically possible solution can not be derived, we find a solution in which the direction of the connecting link coincides with the goal direction (see Fig.4).

$$
\begin{aligned}
\mathbf{g}_i^w = \;& \mathbf{T}^b \mathbf{R}^b(0, R_y{}^b, R_x{}^b)\mathbf{R}^{b'}(R_z{}^{b'}, 0, 0) \\
& \mathbf{T}^{l_1}\mathbf{R}^{l_1}(R_z{}^{l_1}, R_y{}^{l_1}, 0)\mathbf{R}^{l_1'}(R_z{}^{l_1'}, 0, R_x{}^{l_1'}) \quad (4)\\
& \mathbf{T}^{l_2}\mathbf{R}^{l_2}(0, 0, R_x{}^{l_2})\;\mathbf{t}^e
\end{aligned}
$$

where $\mathbf{g}_i^w = (g_x^w, g_y^w, g_z^w, 1)^T$ is a goal vector($i = 1, \cdots, 4$); $\mathbf{T}^b$, $\mathbf{R}^b$ and $\mathbf{R}^{b'}$ are matrices representing the body pose; $\mathbf{T}^{l_1}$, $\mathbf{R}^{l_1}$ $|\mathbf{R}^{l_1'}$, $\mathbf{T}^{l_2}$ and $\mathbf{R}^{l_2}$ are pose matrices related to link 1 (L1 in Fig.3)and link 2 (L2) respectively; $\mathbf{t}^e$ is a translation vector related to the end-effector position of L2.

Here, some rotation elements are represented in two matrices—$\mathbf{R}^{l_1}$ and $\mathbf{R}^{l_1'}$ of L1, for example. We have divided the original rotation matrix into two matrices to simplify analytical solution of our inverse kinematics. In $\mathbf{R}$, R_z, R_y, R_x represent roll, pitch, and yaw angle respectively.

4.1 Analytical Solution Using Real-time Inverse Kinematics

Analytical solution of the inverse kinematics previously mentioned is as follows[1]:

$$
R_y{}^{l_1} = -\arccos\left(\frac{g_z - T_z{}^{l_1}}{\|\mathbf{g} - \mathbf{T}^{l_1}\|}\right) \quad (5)
$$

$$
R_z{}^{l_1} = -\arctan\left(\frac{g_y - T_y{}^{l_1}}{g_x - T_x{}^{l_1}}\right) \quad (6)
$$

$$
R_x{}^{l_1'} = \arccos\left(\frac{L_1{}^2 + L^2 - L_2{}^2}{2L_1 L}\right) \quad (7)
$$

$$
\begin{aligned}
R_x{}^{l_2} &= \arccos\left(\frac{L_1{}^2 + L_2{}^2 - L^2}{2L_1 L_2}\right) - \pi \quad (8)\\
&\quad (|L_1 - L_2| \le L \le L_1 + L_2)
\end{aligned}
$$

where L_1, L_2, L are the lengths of link 1, link 2, and the distance between the origin of link 1 and the goal position.

4.2 Estimation of Torso Posture

Torso posture consists of two elements, the axis of the torso and the pan angle around the axis. The axis of the torso is an axis connecting the centers of gravity of a head blob and a torso blob and is defined as follows:

$$
R_x{}^b = -\arcsin\left(\frac{T_y{}^h - T_y{}^b}{\|\mathbf{T}^h - \mathbf{T}^b\|}\right) \quad (9)
$$

$$
R_y{}^b = -\arctan\left(\frac{T_x{}^h - T_x{}^b}{T_z{}^h - T_z{}^b}\right) \quad (10)
$$

The pan angle (i.e. human body direction), is difficult to estimate correctly from perception results, or blobs. However, since we use multiple cameras, we can estimate the body pan angle for a variety of body postures. We estimate the pan angle based on the direction that both feet point, assuming that both feet touch the ground plane or that they are very close to the ground plane:

$$
R_z{}^{b'} = -\arctan\left(\frac{T_y{}^{rf} - T_y{}^{lf}}{T_x{}^{rf} - T_x{}^{lf}}\right) \quad (11)
$$

The characteristics of our method can be summarized as follows:

- Since only two-link inverse kinematics is solved, it can be used for real-time pose estimation of human bodies.

- Parameters which are not represented explicitly in the solution of the inverse kinematics, such as the pitch of elbow ($R_z{}^{l_1'}$) (we call it the *characteristic angle*), can be used to control precise human body pose if necessary[2]

5 Implementation and Experiments

5.1 Characteristic Angle Estimation from Real Motion Capture Data

In order to investigate features of the characteristic angle $R_z{}^{l_1'}$, we measured its real angle for various arm and leg directions using a marker-based motion capture system[10], in which 4 markers are attached to

---

[1] All the coordinates here are represented in the local coordinate system of the torso.

[2] In this paper, we have used a pre-acquired constant value based on measurements of limb postures using another motion capture device. See5.1.

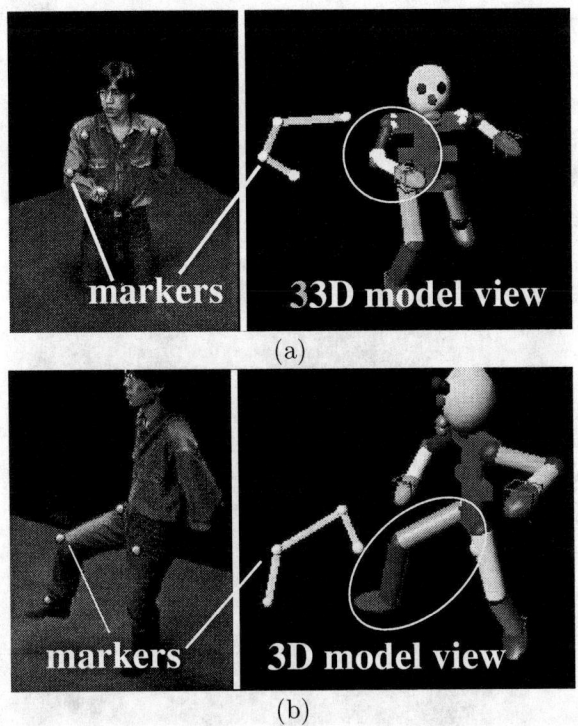

(a)

(b)

Figure 5: Samples of input images and estimated 3-D model views:(a)right arm,(b)right leg.

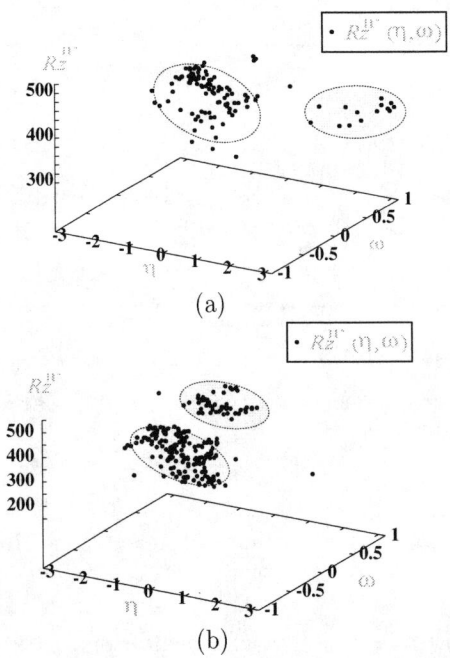

(a)

(b)

Figure 6: plots of characteristic angle: (a)right arm, (b)right leg.

both shoulders(hips), the elbows(knees) and the hands(feet) of the right arm(leg). Fig.5(a)(b) shows samples of input images and of reconstructed 3-D model views, which poses are estimated by our IK method. Fig.7 shows 3-D position trajectories of a right elbow and a right knee. Fig.6 shows plots of the characteristic angle $R_z^{l'_1}$ with various arm and leg directions (latitude η and longitude ω, see Fig.4). From this result, we can suppose that the characteristic angles can be approximated by constant values. In fact, for example, the errors between the elbow positions of the human body generated referring to measured angles and ones generated with a constant angle, which is a mean value of the measured angles, 16[deg] in (a), are small enough to reconstruct natural poses of the model. In case of (b), or in case of the right leg, the mean value of the characteristic angles is 236[deg].

5.2 Real-time Interaction between Human and Virtual Environment

Here, the human figure motion generator mentioned above is applied to *Visually Guided 3D Animation*, or a real-time (video-rate) online interaction system in a virtual space. The example shown in Fig.8 shows a user, or a target object, which is visualized as an *avatar* in the virtual space, kicking a virtual soccer ball. Its interaction is simply realized by detecting collision of the ball and the body parts and by simulating rebound of the ball from the body. In this case, a

delay of about 0.2 second is inevitable because of the latency of the pipelined implementation. Therefore, in these shots(Fig.8), the avatar posture is slightly different from that of user.

6 Conclusions

In this paper, we have shown a real-time human motion capturing without special marker-sensors. We have adopted multi-view fusion and inverse kinematics to realize full-body motion analysis from a limited number of perceptual cues. Since the system works in real-time and online, it can be applied to various *real-virtual* applications such as smart man-machine 3-D interaction. In future work, we will achieve more natural motion of the human model by employing emotional and dynamical filtering.

Acknowledgments

This work has been partly supported by "Cooperative Distributed Vision for Dynamic Three Dimensional Scene Understanding (CDV)" project (JSPS-RFTF96P00501, Research for the Future Program, the Japan Society for the Promotion of Science).

References

[1] C.Wren, A.Azarbayejani, T.Darrell, A.Pentland, "Pfinder: Real-Time Tracking of the Human Body", *IEEE Transactions on Pattern Analysis and Machine Intelligence*, Vol.19, No.7, pp.780-785, 1997.

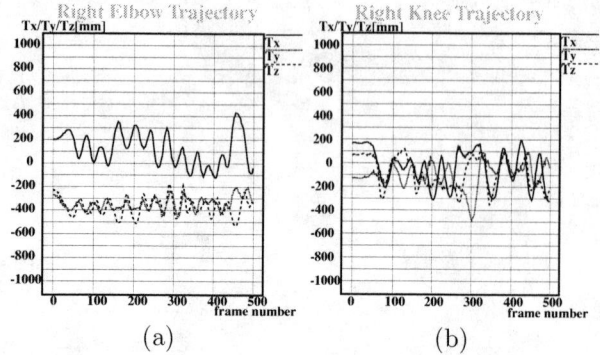

(a)   (b)

Figure 7: 3-D position trajectory: (a)right arm, (b)right leg.

[2] C.Bregler, "Learning and Recognizing Human Dynamics in Video Sequences", *in Computer Vision and Pattern Recognition*, pp.568–574, 1997.

[3] M.Etoh, Y.Shirai, "Segmentation and 2D Motion Estimation by Regioịn Fragments", *in International Conference on Computer Vision*, pp.192–199, 1993.

[4] C.Wren, A.Pentland, "Understanding Purposeful Human Motion", *in Fourth IEEE International conference on Automatic Face and Gesture Recognition*, 2000.

[5] A.Witkin, M.Gleicher, W.Welch, "Interactive Dynamics", *in ACM SIGGRAPH*, Vol.24, no.2, pp.11–21, 1990.

[6] Y.Okamoto and R.Cipolla and H.Kazama and Y.Kuno, "Human Interface System Using Qualitative Visual Motion Interpretation", *IEICE*, Vol.J76-D-II, No.8, pp.1813–1821, 1993.

[7] Myrinet. http://www.myricom.com

[8] D. Arita, N. Tsuruta and R. Taniguchi, "Real-time parallel video image processing on PC-cluster", *Parallel and Distributed Methods for Image Processing II, Proceedings of SPIE*, Vol.3452, pp.23–32, 1998.

[9] J. Zhao and N. Badler: Inverse Kinematics positioning using nonlinear programming for highly articulated figures, Transactions on Computer Graphics, Vol.13, No.4, pp.313–336, 1994.

[10] S.Yonemoto, N.Tsuruta, and R.Taniguchi: "A Real-time Motion Capture System with Multiple Camera Fusion", Proc. ICIAP'99, pp.600-605, 1999.

Figure 8: Online demo shots: (a)*real-virtual soccer scene*, (b)zoom in, (c)6 blob representation. Note that these shots were taken at different times.

Human Motion from Active Contours

Jane Wilhelms, Allen Van Gelder, Leon Atkinson-Derman, Alison Luo
Computer Science Dept., University of California, Santa Cruz, CA 95064
wilhelms,avg,ljderman,alison@cse.ucsc.edu

Abstract

We describe an approach for extracting three-dimensional articulated motion from unrestricted monocular video sequences. We combine feature extraction methods based on active contours with interactive adjustment. An articulated model is interactively aligned with the image in selected anchor frames. Active contour points are anchored to model segments in these frames. Occluded points are detected using object geometry and do not participate in edge tracking. Model joints are automatically adjusted in other frames to align with active contour points. The combination of interactive and automatic adjustment allows extraction of arbitrarily complex movements.

1 Introduction

Although the last twenty years have seen many interesting computer graphics techniques for high-level, constrained control of articulated body motion, it is still not possible to reliably generate truly realistic human and animal motion. Trained animators can do this using keyframing, but they are few in number and expensive. A successful alternative is *motion capture*, where the subject motion is measured and duplicated generally in a studio setting. Motion capture generally occurs in a studio setting where either multiple cameras or body-fixed sensors are used.

However, videos record human and animal motion of greater variety than is encountered in studios. Children playing in the surf, deer leaping, and cats hunting are examples of motion not amenable to extraction by multiple cameras or wearing special apparatus. Occlusion, rapid movement, out-of-plane rotations, and ambiguities hamper automatic approaches. We are exploring methods that combine computer vision techniques with interactive manipulation to extract arbitrary motion from video.

Our long-term goal is to develop a motion library that encodes movement so that it can be reused in different environments and with different creatures. Motions that we extract from video are approximate, but can provide a start-

ing point for such encodings.

2 Background

The computer vision methods most applicable to our work are model-based techniques for extracting articulated body motion from monocular image sequences. The review of Aggarwal *et.al* is an excellent summary of applicable research [1]. Low-level image processing techniques may be used to prepare the images. While features take various forms, *active contours*, or *snakes* [5, 11] are the basis for our work. The model may be a stick-figure or be volumetric [4, 6, 15]. It may be interactively created by the user [2], or generated automatically during extraction [10]. Tracking is achieved by minimizing the error between the model and image [7, 16]. Matching a 2D model with the image may be followed by a separate 3D reconstruction [13], or the two may be combined, with joint and image constraints treated together [8]. Results using optical flow and probabilistic methods are encouraging [9, 14].

3 Method

We use *active contours* [11] as a feature representation that can both automatically react to the image and be easily manipulated by the user. Image processing techniques such as blurring, edge detection, and intensity mapping can be used initially to clarify image content. The user then designs active contours in selected frames; those in intervening frames can be created automatically or with user aid. The 3D model is interactively positioned in selected anchor frames and contour vertices are anchored to the model there. Kinematic adjustment automatically brings the model into alignment with contours in other frames.

3.1 Definitions

We start with a three-dimensional articulated model created interactively using our *zoo* creature modeling and animation software [3, 12, 17]. Articulations are *joints*, and body regions between joints are *segments*. Our standard human model has 70 individual segments, though not all need

155

be involved in the extraction. Segments can be grouped into super-segments, [12], allowing many contiguous segments to be controlled by a few parameters. Each segment is defined in a local coordinate system relative to its parent segment, with the root segment defined in the world. The root segment has six degrees of freedom (rotations and translations) but others only rotate.

Images are digitized from video and texture-mapped onto a movable plane. The world Z-axis is perpendicular to the image plane.

We call a sequence of active contour points that act together a *fauna snake* or *fsnake* to indicate its existence both within the image and as part of the model. Many fsnakes may act on one model. They generally exist in multiple frames; there is a one-to-one relationship between points in different frames.

$\mathbf{P_{i,j}}$ refers to the i-th contour point in the j-th frame. The frame subscript is omitted when it is clear by context. Vectors are given in boldface and scalars in italics. Some calculations are done with three-dimensional vectors, while others use a projection of the vectors onto the image plane. The superscript **pw** (e.g., $\mathbf{P_{i,j}^{pw}}$) indicates a world-space vector that is projected so that its Z-coordinate is zero. Vector components are written as $[V_x, V_y]$.

Figure 1 illustrates the process we describe by showing a two-segment model being matched to an elliptical image figure. In the top diagram, the user positioned the model segments (represented by their labeled coordinate axes) on the image figure and created a three-point fsnake nearby. After automatic anchoring, each contour point of the fsnake is attached to a nearby segment longitudinal axis and this position is stored as an anchor $\mathbf{A}$ in the local segment frame. In an anchor frame, each contour point $\mathbf{P_{i,j}} = \mathbf{A_{i,j}}$.

In the middle diagram showing frame 1, the image figure has moved and the contour points have tracked it, so that they are no longer aligned with their respective anchors. In the lowest diagram, still frame 1, the model has been automatically transformed to align the virtual anchors and contour points. Segment a (the root) has translated and rotated, and segment b has rotated. The following sections describe how this process occurs.

3.2 Creating the Fauna Snakes

The user initially positions the fsnakes in selected frames by picking points on the image. Fsnakes in intervening frames can be created by copying from another frame, by interpolating between fsnakes already created, or by extrapolating positions in two previous frames as an estimate of velocity. An fsnake may be treated as rigid (e.g., representing a single limb) or non-rigid. Rigid fsnakes can be interpolated as a rigid body using translation and rotation, while with non-rigid fsnakes, points are interpolated individually.

The user can also adjust an fsnake in any frame either as a rigid body or by moving individual points.

Within a frame, contour points can be *snapped* into position based upon a weighted combination of internal forces (*length* and *angle*) and external forces (*intensity, hue* and *gravity*). The user can control the weighting. The intensity force may be calculated all along an edge, while the others are just calculated at contour points. During snapping, contour points move at most one pixel per iteration, then forces are recalculated.

Contour points in certain frames are special, in that information such as hue, edge length, and angles between edges, are stored and used to move contour points in other frames. Further details are available [3].

3.2.1 Intensity, Hue and Gravity Forces

The intensity force is generally the most important, as it drives contour points toward an edge. The intensity $e(x, y)$ is a bilinear function of the intensities $e_{i,j}$ at the four image

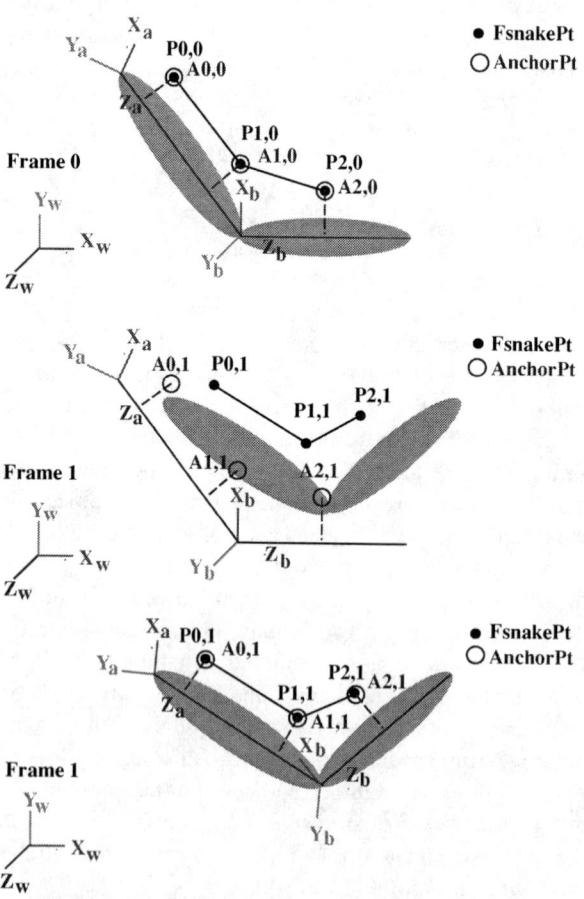

Figure 1. A two-segment model tracks a figure in an image sequence. See Section 3.1.

156

pixels surrounding the point (x, y). For x and y between 0 and 1:

$$e(x, y) = (1 - y)((1 - x)e_{0,0} + xe_{1,0})$$
$$+y((1 - x)e_{0,1} + xe_{1,1})$$

The intensity force is the gradient, $\mathbf{F}_{int}(x, y) = \nabla_{x,y}e(x, y)$.

For the hue force $\mathbf{F}_{hue}$, the $e_{i,j}$ values above are substituted with the difference between the corner hue values and the initial hue at a contour point when it is first created. The hue force drives the contour point toward the most similar hue. Gravity forces $\mathbf{F}_{gravity}$ encourage the contour to go in a particular user-specified direction.

3.2.2 Length- and Angle-Preserving Forces

These forces preserve the original contour shape. The angle-preserving force $\mathbf{F}_{ang}$ for the angle between three contour points $\mathbf{P_{i-1}}, \mathbf{P_i}$ and $\mathbf{P_{i+1}}$ is found by taking the difference α between the angle when the contour was first created and that found at present. Half the force is applied as a torque to one edge and half to the other.

$$\mathbf{F}_{ang}(\mathbf{P_{i-1}}) = 0.25\alpha[-(\mathbf{P_i} - \mathbf{P_{i-1}})_y, (\mathbf{P_i} - \mathbf{P_{i-1}})_x)]$$
$$\mathbf{F}_{ang}(\mathbf{P_{i+1}}) = 0.25\alpha[-(\mathbf{P_{i+1}} - \mathbf{P_i})_y, (\mathbf{P_{i+1}} - \mathbf{P_i})_x)]$$
$$\mathbf{F}_{ang}(\mathbf{P_i}) = -0.25\alpha[-(\mathbf{P_i} - \mathbf{P_{i-1}})_y, (\mathbf{P_i} - \mathbf{P_{i-1}})_x)]$$
$$-0.25\alpha[-(\mathbf{P_{i+1}} - \mathbf{P_i})_y, (\mathbf{P_{i+1}} - \mathbf{P_i})_x)]$$

The length-preserving force $\mathbf{F}_{length}$ is the difference between the initial edge length (in the initial frame) and the present length, applied in equal and opposite directions at the contour points defining an edge. It may maintain absolute length, or relative length between neighboring edges.

3.3 Model to Image Alignment

The initial alignment is done interactively. First the user must find a correct segment size by adjusting the model to the image in appropriate frames. Next the model is positioned in *model key frames* by translating the root segment and rotating joints. The use of super-segments and inverse kinematics simplifies the task [12].

Model key frames are selected because they clearly show a position, or because they define important changes in motion, such as the beginning and end of an out-of-plane trunk rotation. The initial position of the model in other frames can be an interpolation of the position in these model key frames. This is useful, because the interpolated motion includes rotations about longitudinal segment axes and out of the image plane. These motions are hard to track.

3.4 Anchoring Fauna Snakes

Contour points are anchored, in *anchor frames*, either to the segment to which they are nearest, or to a designated segment. If anchoring is to any segment, the nearest distance between a contour point and the projection of the segment longitudinal axis onto the image plane is found. The anchor point $\mathbf{A}$ receives the X and Y coordinates of its contour point, but the Z value is that of the nearest location on the longitudinal axis of the anchor segment. The anchor is transformed into the local coordinate system of the anchor segment, and it moves with the segment in all future frames, until a new anchor frame in encountered.

While fsnake point $\mathbf{P}$ and anchor point $\mathbf{A}$ originally project to the same position on the image plane in the anchor frame, their behavior in other frames is different (see Figure 1). $\mathbf{P}$ is fundamentally a world space position and may change in other frames to track the image figure. $\mathbf{A}$, however, is fundamentally a fixed local position in the anchor segment coordinate frame. Its world space position changes when the model segments translate and rotate.

The model must be appropriately aligned with the image figure in anchor frames. It is useful to re-anchor whenever the fsnake is failing to produce the desired motion in the model.

3.5 Automatically Repositioning the Model

All that came before is to make possible the automatic adjustment of the model in future frames so that the error between contour points and their anchors is minimized by some criteria. Our approach is a *kinematic* one, based only on relative positions; although it is natural to speak of adjustments as due to forces and torques, they are virtual.

The user has discretion over the kind of change that an fsnake can cause to the model, and also over what parts of the model are affected by it. Each contour point may cause a *translation*, an *image-plane rotation*, and/or an *out-of-plane rotation*. Each segment of the body may be affected by: (1) only contour points anchored to that segment; (2) by contour points anchored to the chain of joints to which it belongs (the super-segment); or (3) by any contour points distal to that segment in the model hierarchy. The latter two cause an *inverse kinematic* adjustment to a chain of segments, where a chain of joints is repositioned.

Because distal segment positions are affected by proximal changes, adjustments must be applied recursively from the root of the model outwards. Translations are applied first; then, for each segment, in-plane rotation followed by out-of-plane rotation.

3.5.1 Translation

Translational fauna snakes apply a position change to the model root segment parallel to the image plane. The translation is calculated using the difference between the projected positions in world space of each contour point $\mathbf{P_i^{pw}}$ and its anchor $\mathbf{A_i^{pw}}$. The total translation change of the root segment from n translate fsnake points in frame j is:

$$\mathbf{F_j} = \sum_{i=1}^{n} \frac{\mathbf{P_{i,j}^{pw}} - \mathbf{A_{i,j}^{pw}}}{n}$$

3.5.2 Image-Plane Rotation

Image plane rotations occur around the Z-axis of world space, perpendicular to the image plane. (Rotations are transformed into local segments space for application.) Some set of n contour points (Section 3.5) exert virtual torques on each segment. The torque due to contour point $\mathbf{P}_i$ on segment b around its origin $\mathbf{O_b}$ depends on the angle θ_i between two vectors projected onto the image plane: the vector from the segment origin to the anchor point in world space, which is $\mathbf{A_i^{pw}} - \mathbf{O_b^{pw}}$, and the vector from the segment origin to the contour point in world space, which is $\mathbf{P_i^{pw}} - \mathbf{O_b^{pw}}$ (see Figure 2). The angle θ_i is weighted by the relative squared distance of the projected anchor $\mathbf{A_i^{pw}}$ from $\mathbf{O_b^{pw}}$. Letting r_i denote the actual projected distance, the actual angular change θ_b applied to segment b due to n contour points is then:

$$\theta_b = \frac{\sum_{i=1}^{n} \theta_i r_i^2}{\sum_{i=1}^{n} r_i^2}$$

The axis of rotation is the world Z-axis. However, because rotations must be applied in the local segment space, the world Z-axis must be transformed to local segment space before the rotation is applied.

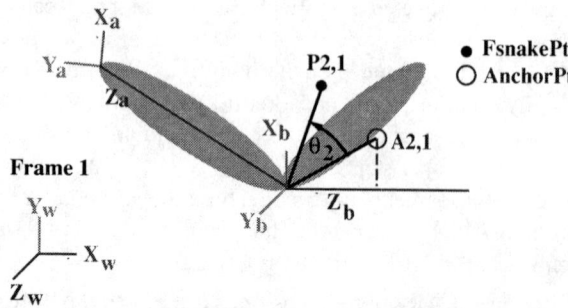

Figure 2. θ_2 is the image plane correction angle due to $\mathbf{P_{2,1}}$ and $\mathbf{A_{2,1}}$.

3.5.3 Out-of-Plane Rotation

If part of the image figure is rotating out of the image plane, projected contour points tracking it will move closer or farther from the point about which rotation occurs. This is mimicked in the model by rotating the appropriate segment about an axis perpendicular to both the segment longitudinal axis and the image plane normal. This rotation will not affect the projected direction of the longitudinal segment axis.

Calculations are done in world space, but applied in local segment space. To simplify the mathematics, we effectively translate the segment to the world origin and rotate around the world Z-axis to place the longitudinal segment axis in the world $Y = 0$ plane; call this rotation angle α. After computing and performing the out-of-plane rotation, we invert the rotation by α and the translation. In the new frame rotated by α, the out-of-plane rotation is simply a rotation around world Y by some angle ϕ and the error for each contour point is simply the difference in X-value between the point and its anchor. Let $\mathbf{P}_i^{\alpha w}$ and $\mathbf{A}_i^{\alpha w}$ be the positions of the contour point i and its anchor after rotation by α. Let $\mathbf{A}_i^{\alpha w}(\phi)$ be the anchor position after a further rotation by ϕ around the Y-axis. Subscripts of x, y, and z denote the X, Y, and Z components of vectors. Thus $A_{ix}^{\alpha w}(\phi) = A_{ix}^{\alpha w} \cos \phi + A_{iz}^{\alpha w} \sin \phi$.

To find the solution angle ϕ, we (approximately) minimize the squared error E, which is the sum over i of the squared distance in X between the i-th contour point and its anchor, in the rotated frame. We (approximately) find the ϕ for which $\partial E / \partial \phi = 0$.

$$E = \sum_i \left(P_{ix}^{\alpha w} - A_{ix}^{\alpha w}(\phi) \right)^2$$

$$\frac{\partial E}{\partial \phi} = \sum_i 2 \left(A_{ix}^{\alpha w}(\phi) - P_{ix}^{\alpha w} \right) \frac{\partial A_{ix}^{\alpha w}(\phi)}{\partial \phi}$$

Dropping higher-order terms in $\tan \phi$, the solution is

$$\tan \phi = \frac{\sum_i A_{iz}^{\alpha w}(P_{ix}^{\alpha w} - A_{ix}^{\alpha w})}{\sum_i \left(A_{ix}^{\alpha w}(P_{ix}^{\alpha w} - A_{ix}^{\alpha w}) + (A_{iz}^{\alpha w})^2 \right)}$$

3.6 Occlusion and Camera Motion

A virtual anatomy can be rendered with our model [17] and used to estimate when contour points become occluded. (So far, we have only tested this with ellipsoidal approximations to limbs.) Because contour anchor points have a world Z value determined by their anchor segment, they become hidden when another part of the anatomy is rendered in front. By giving the anchors a known unique color and checking the color at the projected anchor position in the frame buffer, visibility can be detected. When the anchor is invisible, forces are not calculated for the contour

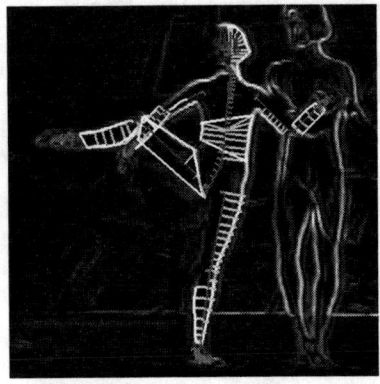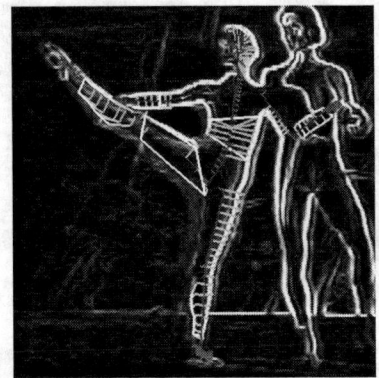

Figure 3. Three Frames from Dance Figure.

point, though it is still influenced by the neighboring points. Although contour points and their anchors do not project to the same position, they are generally close enough that this approach improves tracking.

Our rudimentary approach to removing camera motion is to treat the world coordinate system as an independent segment. Contour points are anchored to this segment in the normal manner and changes in world space are produced. As movement of the world, in which the image plane and the human model exist, is the inverse of the movement of the camera, this motion can be tracked and removed from the human model. In terms of a motion library, other methods of separating the human motion from the world can be applied, such as calculating the position of a known stationary body part or calculating the distance traveled in a stride.

4   Results

We experimented on two sequences: a simple dance movement consisting of a largely planar motion, and the complicated cavorting about of a child in the surf. Color images and animations can be seen on our web site: www.cse.ucsc.edu/~wilhelms/fauna.

The dancer was nicely distinguishable against the simpler background using a combination of median filter, edge detection, greyscale, and intensity adjustment (see Figure 3). The lines perpendicular to the segment axes show the nearest points of contour point anchors to their anchor segments. Eleven fauna snakes were used, and automatic tracking worked fairly well except on the lower arms, whose edges were occluded or unclear in some frames. Only a single initial key position was used for the model. Only the neck fauna snake used inverse kinematics. Fsnakes were applied in 15 frames of 150 frames.

More challenging was the child in the surf (Figure 4). The background was too complex for accurate tracking in most frames, and the user had to guess what was happening when the boy went behind the man. Every other frame of the video was used, for a total of 64 frames. Four complete model positions were originally keyframed, and parts of the model were keyframed further when necessary. Only five fsnakes were used: one translated and oriented the trunk, two moved the left leg, one moved the right leg and one the left arm using inverse kinematics. The right arm was left as keyframed, and missed much of the complexity of the motion. The results shown, from model creation to stored animation, took one afternoon.

We find that when automatic tracking is problematic, creating a different fauna snake for each segment works best. In this case, the fsnake can be treated as rigid and adjusted easily when tracking becomes incorrect, a matter of a few seconds per frame. Further, the motion can be extracted gradually outward from the body root, adding distal fsnakes as needed. The motion extracted can be quite noisy, and additional filtering is necessary. We can apply a mean filter to the captured motion curves (position versus time) at any time during or after extraction.

5   Discussion and Conclusions

While image-plane rotations proved fast and reliable, out-of-plane rotations are more problematic, and need to be used with discretion, because the tendency of contour points to slide along an edge makes angle calculation inaccurate. Rotations along segment longitudinal axes are not explicitly dealt with at all, although keyframing the model in three dimensions produces a reasonable longitudinal rotation in in-between frames, and this remains after adjustment by the contours. No joint limits are applied; we will address that in future work.

We found it possible to catch a usable approximation of complex motion in a reasonable time. Thus we feel that a combination of computer vision and interactive methods can be used to extract articulated motion in cases where completely automatic methods fail.

Figure 4. Boy in the surf: Top shows fsnakes, bottom ellipsoidal model. (Right arm not tracked.)

Acknowledgments: This research was supported by NSF Grants CCR-9503829, CCR-9972464 and CDA-9724237.

References

[1] J. K. Aggarwal, Q. Cai, W. Liao, and B. Sabata. Nonrigid motion analysis: Articulated and elastic motion. *Computer Vision and Image Understanding*, 70(2):142–156, 1998.

[2] K. Akita. Image sequence analysis of real world human motion. *Pattern Recognition*, 17(1):73–83, 1984.

[3] L. Atkinson-Derman. Tracking on the wild side – using active contours to track fauna in noisy image sequences. Master's thesis, UC Santa Cruz, CA 95064, June 2000.

[4] A. Bharatkumar, K. Daigle, M. Pandy, Q. Cai, and J. Aggarwal. Lower-limb kinematics of human walking with the medial axis transformation. *Proc. IEEE Workshop on Motion of Non-Rigid and Articulated Objects*, pages 70–76, Austin, TX, Nov. 11–12 1994.

[5] A. Blake and M. Isard. *Active Contours*. Springer-Verlag, 1998.

[6] Z. Chen and H. Lee. Knowledge-guided visual perception of 3D human gait from single image sequence. *IEEE Trans. on Systems, Man, and Cybernetics*, 22(2):336–342, 1992.

[7] L. Goncalves, E. D. Bernardo, E. Ursella, and P. Perona. Monocular tracking of the human arm in 3D. *Proc. IEEE Fifth Int'l Conference on Computer Vision*, pages 764–770, Cambridge, Mass., 1995.

[8] Y. Hel-Or and M. Werman. Constraint fusion for recognition and localization of articulated and constrained objects. *Int'l Journal of Computer Vision*, 19(1):5–28, July 1996.

[9] M. Isard and A. Blake. Condensation – conditional density propagation for visual tracking. *Int'l Journal of Computer Vision*, 29(1):5–28, 1998.

[10] I. Kakadiaris, D. Metaxas, and R. Bajcsy. 3D human body model acquisition from multiple views. *Proc. IEEE Workshop on Non-Rigid and Articulated Objects*, pages 618–623, Boston, MA, June 20–23 1995.

[11] M. Kass, A. Witkin, and D. Terzopoulos. Snakes: Active contour models. *Int'l Journal of Computer Vision*, 1(4):321–331, 1988.

[12] J. Lapierre. Matching anatomy to model for articulated body animation. Master's thesis, UC Santa Cruz, CA, Dec. 1999.

[13] D. D. Morris and J. M. Rehg. Singularity analysis for articulated object tracking. *Proc. Computer Vision and Pattern Recognition*, pages 289–296, Santa Barbara, CA, June 23–25 1998.

[14] A. Pentland and B. Horowitz. Recovery of nonrigid motion and structure. *IEEE Trans. on Pattern Analysis and Machine Intelligence*, 13(7), 1991.

[15] F. Perales and J. Torres. A system for human motion matching between synthetic and real images based on a biomechanic graphical model. *Proc. IEEE Workshop on Motion of Non-Rigid and Articulated Objects*, pages 83–88, 1994.

[16] J. Rehg and T. Kanade. Digiteyes: Vision-based hand tracking for human-computer interaction. *Proc. IEEE Workshop on Motion of Non-Rigid and Articulated Objects*, pages 16–22, 1994.

[17] J. Wilhelms and A. Van Gelder. Anatomically based modeling. *Proc. ACM SIGGRAPH*, Aug. 1997.

Hand Shape Estimation Using Image Transition Network

Yasushi Hamada, Nobutaka Shimada and Yoshiaki Shirai
Dept.of Computer-Controlled Mechanical Systems, Osaka University
2-1 Yamadaoka, Suita, Osaka 565-0871, Japan
E-mail:hamada@cv.mech.eng.osaka-u.ac.jp

Abstract

This paper presents a method of hand posture estimation from silhouette images taken by two cameras. First, we extract the silhouette contour for a pair of images. We construct an eigenspace from images of hands with various postures. For effective matching, we define a shape complexity for each image to see how well the shape feature is represented. For a pair of input images, the total matching error is computed by combining the two matching errors according to the shape complexity. Thus the best-matched image is obtained for a pair of images. For rapid processing, we limit the matching candidate by using the constraint on the shape change. The possible shape transition is represented by a transition network. Because the network is hard to build, we apply offline learning, where nodes and links are automatically created by showing examples of hand shape sequences. We show experiments of building the transition networks and the performance of matching using the network.

1 Introduction

Recently image-based human interfaces and understanding the hand gestural languages have attracted increasing attentions as an alternative to traditional input devices like mouses or keyboards. Such attempts previously proposed are approximately divided into two categories.

The first category is the 3-D model-based approach including the model fitting methods[1] and "Estimation by Synthesis(ES)" methods [2][3] which match possible postures generated from a given 3-D shape model and search for the postures best-matched to the input image. While these methods are effective for recognition of arbitrary hand postures, they often require much computation.

The second category directly matches the image features to those of models. The methods of this category[4][5][6][7][8] register the image appearances or the image features in the learning sequences, and then the input sequence is classified into one of the registered sequence. For recognition of a limited set of hand postures, only useful models are registered. Moreover, computation is usually less because 3-D shapes are not estimated.

For recognition of hand shapes in a gesture sequence, however, the first category is more effective because it is able to limit the search space by the constraint of the joint angles or by that of the velocity. The second category, on the other hand, has to try to match every models.

This paper proposes a method of matching a given hand posture just like the second category, while limiting the candidates by a transition network built during a learning phase. While the Hidden Markov Model (HMM) approach has to build sequence models for all gesture sequences, this transition network alone represents the transition of all possible gestures.

First, in this paper, a basic matching method is described. We determine the features for a pair of images to estimate the hand posture. We collect various hand images to make the model of the postures. A silhouette is extracted from each image and the feature vector is computed as a sequence of the distances from the center of the silhouette to the contour points. The eigenvectors are determined from all feature vectors.

For effective matching, we define the shape complexity for each image to see how well the shape feature is represented. For a pair of input images, the total matching error is computed by combining the two matching errors according to the shape complexity. Thus the best-matched image is obtained for a pair of images.

Next, an effective hand posture matching of a gesture sequence is described. For a given application, we may be able to limit the matching candidate by the constraint on the shape change. That is, the next shape is confined to a set of possible models. The possible transition is represented by a transition network. Because the transition network is hard to build, we apply offline learning, where nodes and links are automatically created by showing examples of hand shape sequences. It is important to merge similar nodes in different image sequences so that the transition obtained in a sequence can be used at the similar node in other sequences.

2 Feature Extraction

2.1 Contour Feature

For simplicity, the hand region is assumed to be brighter than the background and the clothes so that the hand region

161

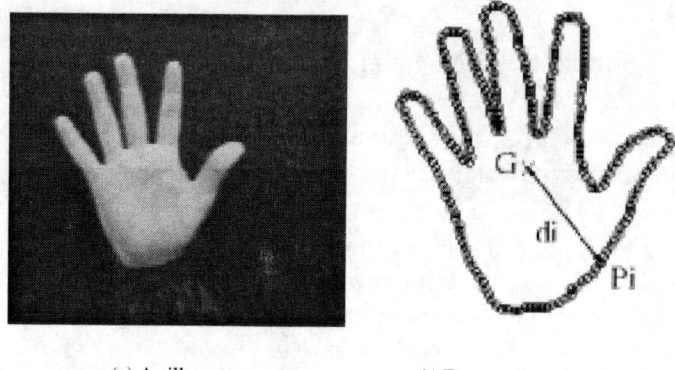

(a) A silhouette   (b) Extracted contour feature

Figure 1. Feature extraction

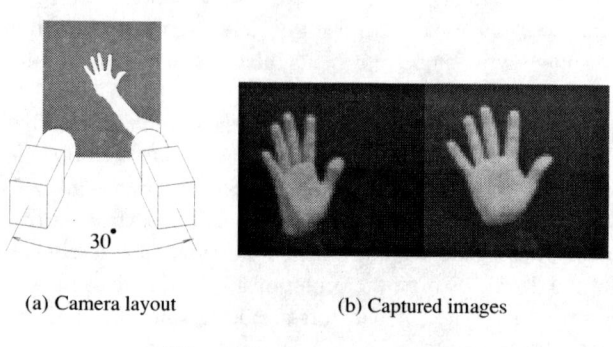

(a) Camera layout   (b) Captured images

Figure 2. Stereo images

 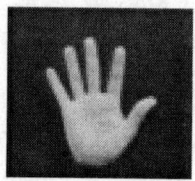

(a) Hand Image 1   (b) Hand Image 2

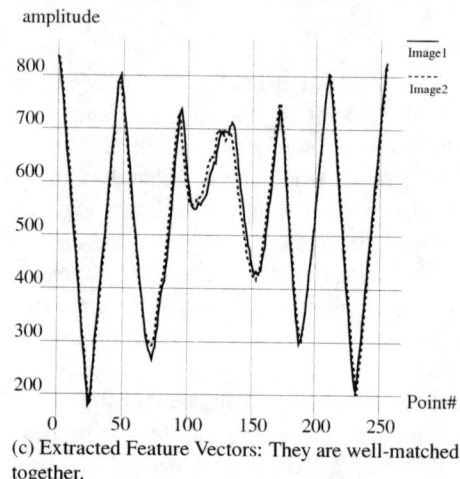

(c) Extracted Feature Vectors: They are well-matched together.

Figure 3. Feature Vectors of Similar Images

is easily obtained (Figure 1(a)).

A pair of hand images are obtained by two cameras fixed laterally in front of the user (Figures 2(a) and 2(b)). For each image of the pair, hand region is extracted and then its area S and center of gravity G are computed. Then 256 points $P_i(i = 1 \cdots 256)$ are sampled on the contour of the region so that they are placed at a constant interval (Figure 1(b)). Scale-normalized distance r_i is obtained by

$$r_i = \frac{d_i}{\sqrt{S}} \tag{1}$$

where d_i donotes the distance between G and P_i. This is the shape feature independent of the translation and scale change.

Because the sequence of features depends on rotation, the realignment of the elements is necessary. We select the most significant peak or valley as the start point of r_i. Then realigned $\{r_i\}$, $\boldsymbol{x} = \{r_{a_1}, \cdots, r_{a_{256}}\}^T$, is obtained as the feature vector. Figure 3 shows extracted feature vectors for two similar hand images.

2.2 Building of Eigenspace

In the offline learning phase, possible hand shapes are registered as the model images. For efficient registration, the eigenspace of the feature vector is constructed. The bases of the eigenspace are computed by selecting k principal eigenvectors $\boldsymbol{E} = [\boldsymbol{e}_1, \cdots, \boldsymbol{e}_k]$ obtained by Principal Component Analysis. The compressed feature vectors $\boldsymbol{g}_n = \boldsymbol{E}^T(\boldsymbol{x}_n - \bar{\boldsymbol{x}})$ $(n = 1, \cdots, M)$ are stored in the database.

In the online shape estimation, scale-normalized distances $\{r_i\}$ are similarly obtained. For normalization of the rotation, we select start point candidates as the significant peaks and valleys. For robust normalization, we select L candidates and evaluate each of them. For the jth candidate $(1 \leq j \leq L)$, feature vector $\boldsymbol{y}_j = \{r_{b_{j1}}, \cdots, r_{b_{j256}}\}^T$ is generated as the jth realigned $\{r_i\}$.

Each $\boldsymbol{y}_j$ is projected into the eigenspace and then the compressed feature vector of the input is computed as

$$\boldsymbol{h}_j = \boldsymbol{E}^T(\boldsymbol{y}_j - \bar{\boldsymbol{x}}).$$

All candidates are matched to the model features to determine the best-matched model.

3 Appearance Matching Using Stereo Images

The basic matching criterion for L feature vectors and the model image n is

$$d_n = \min_{j=1,\cdots,L}(\|\boldsymbol{h}_j - \boldsymbol{g}_n\|) \qquad (2)$$

The best-matched model is determined as

$$d = \min_{n=1,\cdots,M}(d_n) \qquad (3)$$

Because matching is often ambiguous, we use a pair of stereo images. The matching scheme for stereo images is described in this section.

3.1 Matching based on Shape Complexity

Since two input feature vectors are obtained from stereo images, two matching criteria are computed by equation (3). Note that some hand shapes are difficult to discriminate from a single silhouette.

A problem is how to integrate them to determine the best model.

A simple method is to use the average of the two criteria. This method, however, may be influenced by a bad silhouette (which is not suitable to determine a unique shape).

We conjecture that the more complex are the shapes, the more effectively they represent the 3-D hand shape. The complexity of the shape feature is defined as

$$c = \sum_{i=1}^{256} \frac{|r_{i+k} - r_i|}{k} \qquad (4)$$

where k is an experimentally determined constant. $k = 10$ was used in the experiments.

If the complexity of one image is much more than the other, only the former may be used for matching. In general, each of the stereo images is assigned to a weight (w_l, w_r) according to the complexity, and the best model is determined by the weighted average of the two criteria.

Let the complexity of the left and right image be c_l, c_r. The computation of the weight is experimentally determined in the following way

If $\sqrt{c_l^2 + c_r^2} \le t_1$,
   then $w_l = c_l$, $w_r = c_r$.
If $1/t_2 \le c_l/c_r \le t_2$,
   then $w_l = c_l$, $w_r = c_r$.
Otherwise,
   $w_1 = 1, w_2 = 0$ for $c_1 > c_2$.
where t_1 and $t_2 (\ge 1)$ are determined by experiments. In following experiments, $t_1 = 7.84$ and $t_2 = 1.23$ are adopted.

3.2 Result of Experiment

First the eigenspace is built by typical hand images. By experiments, the performance saturates with 12 eigenvectors. Therefore, 12 dimensional eigenspace is used in the following sections.

By a recognition experiment with 260 model images of different hand postures and 74 input images, recognition rate 95.9% was obtained. Examples of input images and the recognition results are shown in Figure 4.

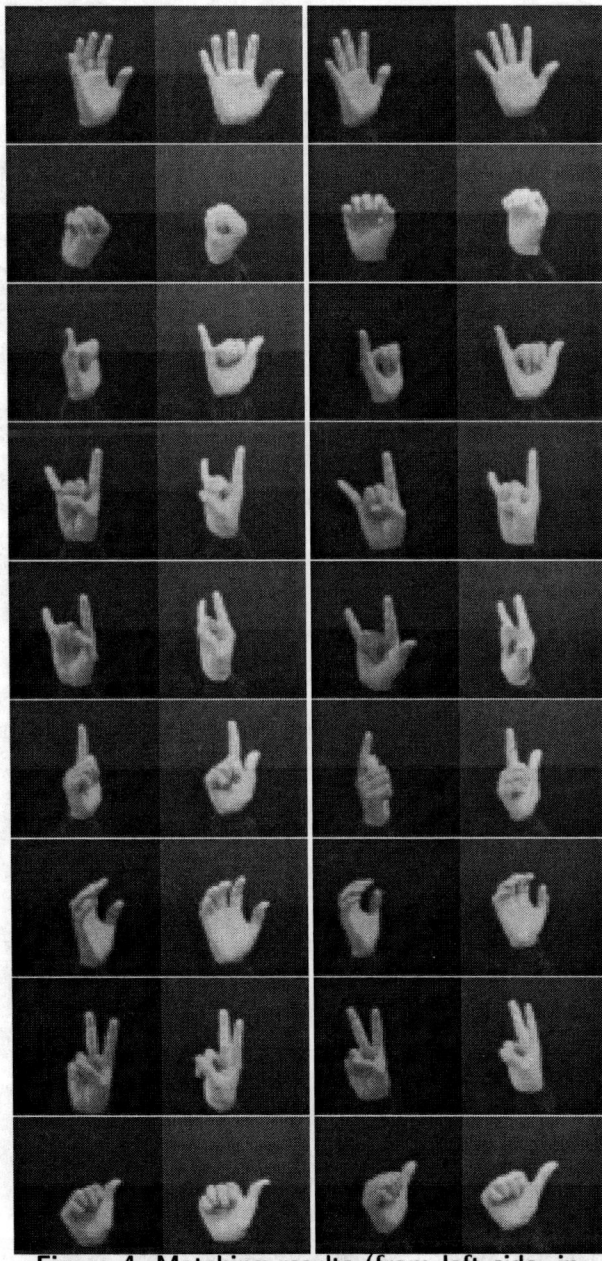

Figure 4. Matching results (from left side, input images(left, right) and matched model images(left and right)

163

4 Transition Network

For recognition of gestures or a hand sign language, a sequence of hand shapes should be recognized. For a given set of gestures, a limited set of shape changes is allowed.

For efficient matching, possible shape changes are stored in a transition network, where nodes represent typical shapes and links represent possible transitions. Generally such a transition network is difficult to build because it takes much efforts to teach all possible transitions.

This section describes a method to build a transition network by showing a limited number of gesture sequences and effective recognition of a shape sequence using the network.

4.1 Building of Transition Network

The transition network is represented in the eigenspace which is built for recognition of hand shapes described in the previous sections.

For learning the network, sample sequences are taken and the network is incrementally built. For a given sample sequence, a sequence of feature vectors is first created in the eigenspace. Each vector is then matched to model nodes using the criterion d described in section 3.

If d is less than a threshold d_{thres}, it is matched to the model node ($d_{thres} = 1.6$ in this paper). If the matched node is the same as the previous node, the hand shape is regarded as the same as the previous one. In this case, no transition takes place.

If the merged node is different from the previous one, the feature vector is merged to it and the link associated to this feature vector is also attached to the model node. By this operation, a new possible transition is automatically created without actual samples. Figure 5 depicts this case.

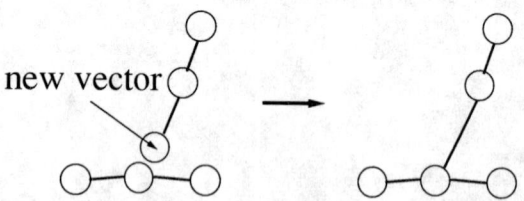

Figure 5. Transition network

If d is greater than d_{thres}, it is regarded as a new shape. Then the new shape becomes a model node.

By repeating this operation for sample sequences, typical hand shapes and possible transitions are represented in the transition network. Note that each node corresponds to the stereo pair of images and the shape feature (in eigenspace).

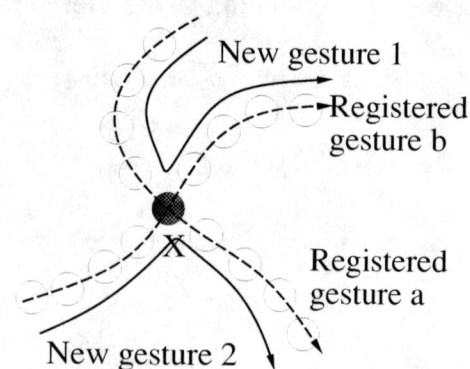

Figure 6. Junction node of transition network

4.2 Shape sequence recognition using Transition Network

In recognition of a shape sequence, the transition network is utilized to find next shape candidates. Given the previous recognition result, the shape candidates are determined as the neighbor of the previous node. Thus the computation cost is much reduced.

Because the junction node (with more than two links) is automatically generated during the learning phase, a new sequence can be tracked. In figure 6, for example, gesture **a** and **b** have been shown and junction X has been generated. If a part of gesture **a** followed by a part of gesture **b** (gesture 1 in the figure) is shown in the recognition phase, it is successfully tracked in the network. Also gesture 2 can be tracked similarly .

5 Experimental Results

We made an experiment of building a transition network and recognizing gesture sequences.

In learning, we prepared 8 kinds of gestures each of which consists of 5 to 35 sample sequences. Each sequence consists of approximately 200 frames. Figure 7 shows 4 kinds of gesture (only 8 typical frames are shown).

In total, 141 sequences and 28200 frames are shown and a transition network with 258 nodes is generated. The number of the node is not very large because many frames correspond to one node.

Figure 8 shows the generated network where the node and the link is represented by a point and a line. Only two principal components in the 12 dimensional eigenspace are shown.

Table 1 shows how many nodes are generated by showing the gestures in the figure.

Next, an experiment with new gesture sequences are performed. Figure 9(a) shows a sequence which consists of multiple kinds of learned gestures. The first part the se-

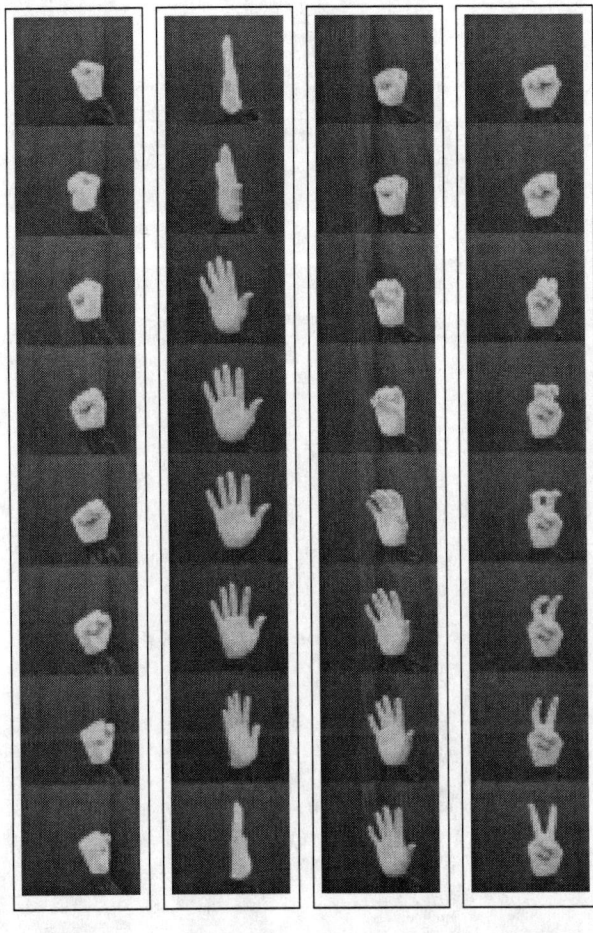

(a) Gesture A (b) Gesture B (c) Gesture C (d) Gesture D

Figure 7. Gesture sequences used for building transition network (left images only. Top of each column is the start of a sequence and the bottom is the end.)

quence is similar to a part of a learned gesture B. Then the sequence shifts to another gesture which is similar to learned gesture C, and shifts to the third gesture D. The result of recognition is shown in Figure 9(b).

Although this sequence is not learned explicitly, the system traced the transition network successfully. Figure 10 shows the trace of gesture A - D shown in Figure 7 in the transition network (thick gray lines), and the recognition result of new gesture sequence (Figure 9(b)) projected on the same network (thick black lines). We can see that the sequence shifts from gesture B through C to D.

We compare the number of matching trials to estimate the efficiency of the proposed method. In the above example, the number of matching for recognition of an image is 258, which means 51600 matching trials are necessary for

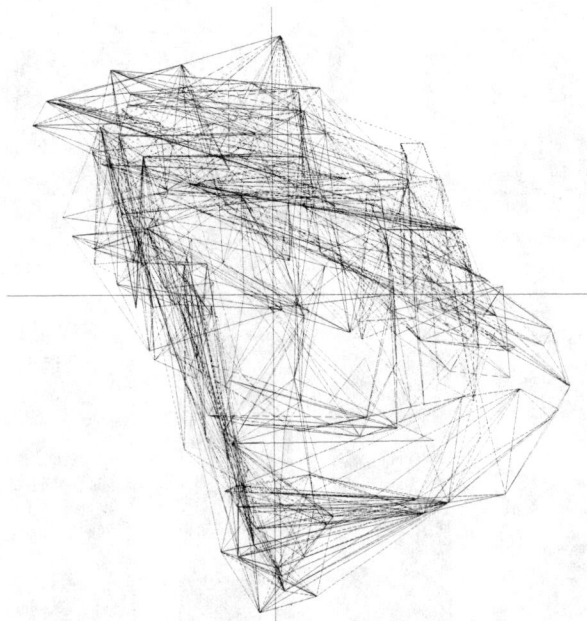

Figure 8. Generated transition network

Table 1. The number of newly generated nodes

Gesture A	3 nodes
Gesture B	49 nodes
Gesture C	16 nodes
Gesture D	2 nodes

recognition of 200 frames.

By using the transition network, the proposed method tries only linked nodes. Because the number of linked nodes is not constant, the number of trials depends on the input sequence. For the above sequence, the number of trials is reduced to 5789 which is 11% of the conventional method.

Now that the average number of links for a node is 8.9 in the generated network. This means that for a random sequence, average number of trials is about 3% of the conventional method.

6 Conclusion

This paper presented a method of the hand posture estimation of gesture sequences from silhouette images taken by two cameras. In the offline learning phase, we construct an eigenspace of image features. In the online recognition phase, the complexity of the left and right image are first evaluated, and the best-matched model is determined by integrating the both matching results on the basis of the complexities.

For efficient recognition of gesture sequences, the shape transition network is proposed. In the learning phase, the

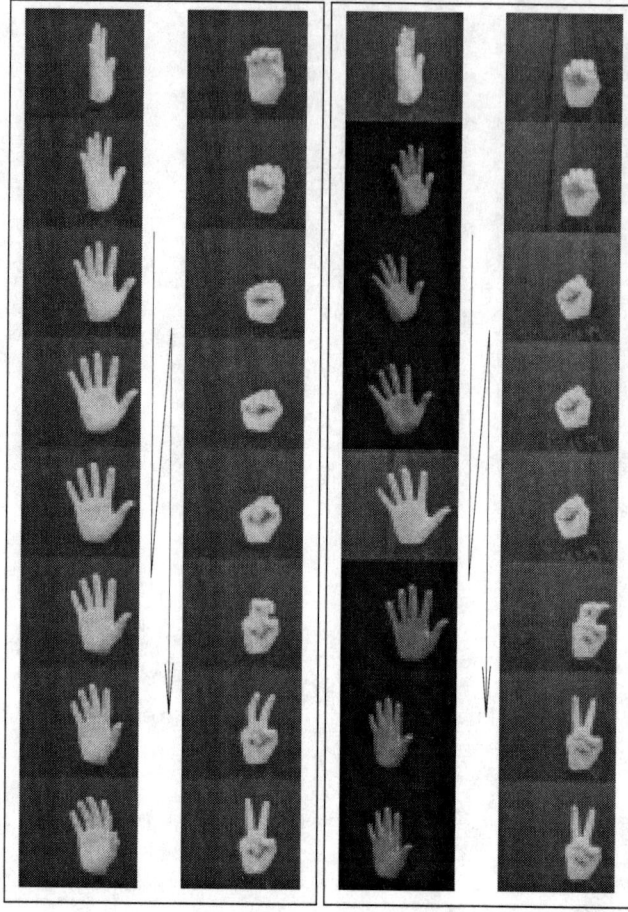

(a) Input test sequence   (b) Matched model

Figure 9. Posture estimation result by shape tracking with transition network (left images only)

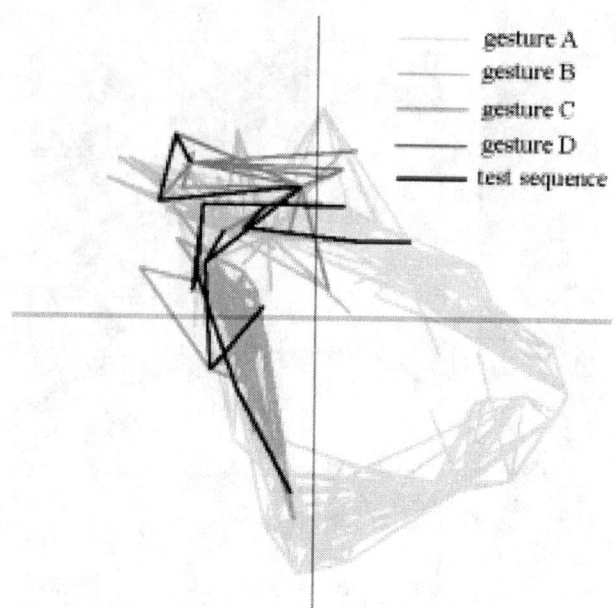

Figure 10. Trace of the test sequence on the transition network

network is automatically generated from the gesture sequences. In the recognition phase, the shape candidates are determined as the neighbor of the previous node in the network. An experiment proved that the computation is reduced to 11% of the conventional method.

The proposed method just recognizes a hand shape. A future work is to recognize a sequence of hand shapes as a meaningful unit such as a word in a sign language.

References

[1] J. M. Rehg and T. Kanade. "Visual Tracking of High DOF Articulated Structures: an Application to Human Hand Tracking". *ECCV'94*, pp. 35–46, 1994.

[2] J. J. Kuch and T. S. Huang. "Virtual Gun: A Vision Based Human Computer Interface Using the Human Hand". In *MVA'94*, pp. 196–199, 1994.

[3] N. Shimada, Y. Shirai, Y. Kuno, and J. Miura. "Hand Gesture Estimation and Model Refinement using Monocular Camera". In *Proc. of 3rd Int. Conf. on Automatic Face and Gesture Recognition*, pp. 268–273, 1998.

[4] Y. Cui and J. Weng. "Learning-based Hand Sign Recognition". *Proc.of Int.Workshop on Automatic Face and Gesture Recognition*, pp. 201–206, 1995.

[5] A. D. Wilson and A. F. Bobick. "Configuration States for the Representation and Recognition of Gesture". *Proc.of Int.Workshop on Automatic Face and Gesture Recognition*, pp. 129–136, 1995.

[6] B. Moghaddam and A. Pentland. "Maximum Likelihood Detection of Faces and Hands". *Proc.of Int.Workshop on Automatic Face and Gesture Recognition*, pp. 122–128, 1995.

[7] T. Nishimura, T. Mukai, and R. Oka. "Spotting Recognition of Human Gestures performed by People from a Single Time-Varying Image". In *Proc. of IROS'97 vol. 2*, pp. 967–972, 1997.

[8] M. J Black and A. D. Jepson. "EigenTracking: Robust Matching and Tracking of Articlated Objects Using a View-Based Representation". *Int.J.of Computer Vision 26(1)*, pp. 63–84, 1998.

Author Index

IEEE
COMPUTER SOCIETY

IEEE Computer Society Publications

The world-renowned IEEE Computer Society publishes, promotes, and distributes a wide variety of authoritative computer science and engineering texts. These books are available from most retail outlets. Visit the Online Catalog, *http://computer.org*, for a list of products.

IEEE Computer Society Proceedings

The IEEE Computer Society also produces and actively promotes the proceedings of more than 141 acclaimed international conferences each year in multimedia formats that include hard and softcover books, CD-ROMs, videos, and on-line publications.

For information on the IEEE Computer Society proceedings, send e-mail to *cs.books@computer.org* or write to Proceedings, IEEE Computer Society, P.O. Box 3014, 10662 Los Vaqueros Circle, Los Alamitos, CA 90720-1314. Telephone +1 714-821-8380. FAX +1 714-761-1784.

Additional information regarding the Computer Society, conferences and proceedings, CD-ROMs, videos, and books can also be accessed from our web site at *http://computer.org/cspress*

Revised 9 November 1999